T0201143

*UNDERSTANDING
SYMMETRICAL
COMPONENTS FOR
POWER SYSTEM
MODELING*

UNDERSTANDING SYMMETRICAL COMPONENTS FOR POWER SYSTEM MODELING

J. C. DAS

WILEY

For general information on our other products and services or for technical support, please contact our Customer Care Department within the United States at (800) 762-2974, outside the United States at (317) 572-3993 or fax (317) 572-4002.

Wiley also publishes its books in a variety of electronic formats. Some content that appears in print may not be available in electronic formats. For more information about Wiley products, visit our web site at www.wiley.com.

Library of Congress Cataloging-in-Publication Data is available.

ISBN: 978-1-119-22685-7

Printed in the United States of America

10 9 8 7 6 5 4 3 2 1

CONTENTS

ABOUT THE AUTHOR

J. C. DAS is an independent consultant at Power System Studies Inc., Snellville, GA. Earlier, he headed the electrical power systems department at Amec Foster Wheeler for the last 30 years. He has varied experience in the utility industry, industrial establishments, hydroelectric generation, and atomic energy. He is responsible for power system studies, including short-circuit, load flow, harmonics, stability, arc-flash hazard, grounding, switching transients, and protective relaying. He conducts courses for continuing education in power systems and has authored or coauthored about 68 technical publications nationally and internationally. He is author of the books:

- *Arc Flash Hazard Analysis and Mitigation*, IEEE Press, 2012.
- *Power System Harmonics and Passive Filter Designs*, IEEE Press, 2015.
- *Transients in Electrical Systems: Analysis, Recognition, and Mitigation*, McGraw-Hill, 2010.
- *Power System Analysis: Short-Circuit Load Flow and Harmonics*, Second Edition, CRC Press, 2011.

These books provide extensive converge, running into more than 3000 pages, and are well received in the technical circles. His interests include power system transients, EMTP simulations, harmonics, passive filter designs, power quality, protection, and relaying. He has published more than 200 electrical power system study reports for his clients.

Mr. Das is a life fellow of Institute of Electrical and Electronics Engineers, IEEE (USA); a member of the IEEE Industry Applications and IEEE Power Engineering societies; a fellow of Institution of Engineering Technology (UK); a life fellow of the Institution of Engineers (India); a member of the Federation of European Engineers (France); a member of CIGRE (France); etc. He is a registered Professional Engineer in the States of Georgia and Oklahoma, a Chartered Engineer (C. Eng.) in the United Kingdom, and a European Engineer (Eur. Ing.) in the European Union (EU). He received meritorious award in engineering, IEEE Pulp and Paper Industry in 2005.

He received MSEE degree from Tulsa University, Tulsa, Oklahoma; and BA (advanced mathematics) and BEE degrees from Punjab University, India.

FOREWORD

THIS BOOK BY J. C. DAS OFFERS AN IN-DEPTH, practical, yet intellectually appealing treatment of symmetrical components not seen since the late Paul M. Anderson's classic, *Analysis of Faulted Power Systems*, which was first published in 1995 by the Wiley-IEEE Press in the Power Engineering Series. The present book leverages the author's well over 30 years of experience in power system studies, and continues in his same tradition of attention to details, which should appeal to those professionals who benefitted from his writing style demonstrated in his four earlier books. The subject is taught at the undergraduate and graduate courses in most universities with a power systems option.

The advent of the symmetrical components concept is due to the Westinghouse electrical engineer Charles LeGeyt Fortescue, who was born in 1876 at York Factory in Manitoba, Canada, who became the first electrical engineer to graduate from Queen's University at Kingston in Ontario, Canada, in 1898. In 1918, Fortescue contributed an 88 page, now classic, remarkable paper by the title "Method of Symmetrical Coordinates Applied to the Solution of Polyphase Networks" in the Transactions of the American Institute of Electrical Engineers (AIEE), one of the two predecessors of present day IEEE. This breakthrough is due to Fortesue's investigations of railway electrification problems which began in 1913. Following the paper's publication, the earlier name "Symmetrical Coordinates" was changed to "Symmetrical Components" and the approach gained in popularity ever since it was disclosed as an indispensable method of dealing with unbalanced three-phase operation problems of electric power systems. A thorough understanding of the application of symmetrical components is required for proper design of electric power protection systems.

Chapter 1 uses matrix algebra to demonstrate the non-uniqueness of symmetrical component transformations. Chapter 2 treats sequence impedances, their networks, and their reduction. Chapters 3 and 4 discuss symmetrical component applications in generating models for transmission lines, cables, synchronous generators, and induction motors. Chapter 3 notes that much of the theoretical underpinnings of the area discussed should be reviewed elsewhere. Prior to discussing three-phase models of two-winding three-phase transformers and conductors, Chapter 5 begins by advising the reader to study this chapter along with Chapter 7. Chapter 6 covers unsymmetrical shunt and series faults and also calculations of overvoltages at the fault plane.

M. E. El-Hawary

PREFACE AND ACKNOWLEDGMENTS

THIS SHORT BOOK consisting of seven chapters attempts to provide a clear understanding of the theory of the symmetrical component transformation and its applications in power system modeling.

Chapter 1 takes a mathematical approach to document that the symmetrical component eigenvectors *are not unique* and one can choose arbitrary vectors meeting the constraints, but these will not be very meaningful in the transformation—thus selection of vectors as they are forms a sound base of the transformation. This is followed by Chapter 2, which details the concepts of sequence impedances, their models, formation of sequence impedance networks and their reduction. Chapters 3 and 4 are devoted to symmetrical component applications in generating the models for transmission lines, cables, synchronous generators, and induction motors. Chapters 5 and 7 are meant to be read together and describe three-phase models and phase-coordinate method of solution where the phase-unbalance in the power system cannot be ignored and symmetrical components cannot be applied. Chapter 6 covers unsymmetrical shunt and series faults and also calculations of overvoltages at the fault point (COG). It has a worked out longhand example to illustrate the complexity of calculations even in a simple electrical distribution system. This is followed with the matrix methods of solution which have been adopted for calculations on digital computers. The author is thankful and appreciates all the cooperation and help received from Ms. Mary Hatcher, Wiley-IEEE and her staff in completing this publication. She rendered similar help and cooperation for the publications of author's other two books by IEEE Press (see Author's profile). An author cannot expect anything better than the help and cooperation rendered by Ms. Mary Hatcher.

The authors special thanks go to Dr. M.E. El-Hawary, Professor of Electrical and Computer Engineering, Dalhousie University, Canada for writing the Foreword to this book. He is a renowned authority on Electrical Power System; the author is grateful to him, and believes that this Foreword adds to the value and the marketability of the book.

J. C. Das

CHAPTER 1

SYMMETRICAL COMPONENTS USING MATRIX METHODS

THE METHOD of symmetrical components was originally proposed by Fortescue in 1918 [1]. We study three-phase balanced systems, by considering these as single-phase system. The current or voltage vectors in a three-phase balanced system are all displaced by 120 electrical degrees from each other. The fundamental texts on electrical circuits [2] derive the equations governing the behavior of three-phase balanced systems. This simplicity of representing a three-phase as a single-phase system is lost for unbalanced systems. The method of symmetrical components has been an important tool for the study of unbalanced three-phase systems, unsymmetrical short-circuit currents, models of rotating machines and transmission lines, etc.

There have been two approaches for the study of symmetrical components:

- A physical description, without going into much mathematical matrix algebra equations.
- A mathematical approach using matrix theory.

This book will cover each of these two approaches to provide a comprehensive understanding. The mathematical approach is adopted in this chapter followed by Chapter 2, which provides some practical concepts and physical significance of symmetrical components. Some publications on symmetrical components are in References [3–6].

It can be mentioned that in the modern age of digital computers, the long-hand calculations using symmetrical components is outdated. See an example of short-circuit calculations in Chapter 6, which is a tedious and lengthy hand calculation for a simple system consisting of five components connected to two buses. In practical power systems, the number of buses can exceed 1000. Today, the systems are modeled with raw input data, and the programs will calculate the sequence components and apply these to derive a result demanded by the problem. Yet, an understanding of symmetrical components is necessary to understand the results from system simulation programs. The reader should understand the limitations of simulation models which are discussed in the chapters to follow.

Understanding Symmetrical Components for Power System Modeling, First Edition. J.C. Das.

1.1 TRANSFORMATIONS

Symmetrical component method is a *transform*. There are number of transformations in electrical engineering, for example, Laplace transform, Fast Fourier transform, Park's transform, Clarke component transform, and the like. There are three steps that are applicable in any transform for the solution of a problem:

- The parameters of the original problem are transformed by the application of the transform to entirely new parameters.
- The solution with the altered parameters is arrived at. The fundamental concept is that the transformed parameters are much easier to solve than the parameters of the original problem.
- Inverse transform is applied to the solved parameters to get to the solution of the original problem.

1.2 CHARACTERISTIC ROOTS, EIGENVALUES, AND EIGENVECTORS

The matrix theory can be applied to understand some fundamental aspects of symmetrical components. The reader must have some knowledge of the matrices as applied to electrical engineering [7, 8], though enough material is provided for continuity of reading.

1.2.1 Definitions

1.2.1.1 Characteristic Matrix
For a square matrix $\bar{A}$, the matrix formed as $|\bar{A} - \lambda \bar{I}|$ is called the *characteristic matrix*. Here λ is a scalar and $\bar{I}$ is a unity matrix.

1.2.1.2 Characteristic Polynomial
The determinant $|\bar{A} - \lambda \bar{I}|$ when expanded gives a polynomial is called the *characteristic polynomial* of matrix $\bar{A}$.

1.2.1.3 Characteristic Equation
The equation $|\bar{A} - \lambda \bar{I}| = 0$ is called the *characteristic equation* of matrix $\bar{A}$.

1.2.1.4 Eigenvalues
The roots of the characteristic equation are called the *characteristic roots* or *eigenvalues*.

1.2.1.5 Eigenvectors, Characteristic Vectors
Each characteristic root λ has a corresponding non-zero vector $\bar{x}$ that satisfies the equation

$$|\bar{A} - \lambda \bar{I}| \, \bar{x} = 0 \qquad (1.1)$$

The non-zero vector $\bar{x}$ is called the *characteristic vector* or *eigenvector*.

Some properties of the eigenvalues are:

- Any square matrix $\bar{A}$ and its transpose $\bar{A}'$ have the same eigenvalues.
- The sum of the eigenvalues of a matrix is equal to the trace of the matrix (the sum of the elements on the principal diagonal is called the trace of the matrix).
- The product of the eigenvalues of the matrix is equal to the determinant of the matrix. If

$$\lambda_1, \lambda_2, \ldots, \lambda_n$$

are the eigenvalues of $\bar{A}$, then the eigenvalues of

$$
\begin{aligned}
k\bar{A} & \text{ are } k\lambda_1, k\lambda_2, \ldots, k\lambda_n \\
\bar{A}^m & \text{ are } \lambda_1^m, \lambda_2^m, \ldots, \lambda_n^m \\
\bar{A}^{-1} & \text{ are } 1/\lambda_1, 1/\lambda_2, \ldots, 1/\lambda_n
\end{aligned}
\tag{1.2}
$$

- Zero is a characteristic root of a matrix, only if the matrix is singular.
- The characteristic roots of a triangular matrix are diagonal elements of the matrix.
- The characteristics roots of a Hermitian matrix are all real.
- The characteristic roots of a real symmetric matrix are all real, as the real symmetric matrix will be Hermitian. A square matrix $\bar{A}$ is called a Hermitian matrix if every i-jth element of the matrix is equal to the conjugate complex j-ith element, that is, the matrix

$$
\begin{vmatrix}
1 & 2+j3 & 3+j \\
2-j3 & 2 & 1-j2 \\
3-j & 1+j2 & 5
\end{vmatrix}
$$

is a Hermitian matrix.

Example 1.1   Find eigenvalues and eigenvectors of matrix

$$
\bar{A} = \begin{vmatrix}
-2 & 2 & -3 \\
2 & 1 & -6 \\
-1 & 2 & 0
\end{vmatrix}
$$

Write the characteristic equation

$$
\bar{A} = \begin{vmatrix}
-2-\lambda & 2 & -3 \\
2 & 1-\lambda & -6 \\
-1 & 2 & 0-\lambda
\end{vmatrix} = 0
$$

Its solution can be shown to be

$$(\lambda+3)(\lambda+3)(\lambda-5) = 0$$

Therefore, the eigenvalues are

$$\lambda = -3, \ -3, \ 5$$

When $\lambda = -3$,

$$\begin{vmatrix} -2+3 & 2 & -3 \\ 2 & 1+3 & -6 \\ -1 & 2 & 3 \end{vmatrix} \begin{vmatrix} x_1 \\ x_2 \\ x_3 \end{vmatrix} = \begin{vmatrix} 0 \\ 0 \\ 0 \end{vmatrix}$$

By manipulation of rows, this can be written as

$$\begin{vmatrix} 1 & 2 & -3 \\ 0 & 0 & 0 \\ 0 & 0 & 0 \end{vmatrix} \begin{vmatrix} x_1 \\ x_2 \\ x_3 \end{vmatrix} = \begin{vmatrix} 0 \\ 0 \\ 0 \end{vmatrix}$$

Therefore,

$$x_1 + 2x_2 - 3x_3 = 0 \tag{1.3}$$

Assume that

$$x_1 = k_1$$
$$x_2 = k_2$$

Then,

$$x_3 = \frac{1}{3}(k_1 + 2k_2)$$

The eigenvectors are

$$\begin{vmatrix} k_1 \\ k_2 \\ \frac{1}{3}(k_1 + 2k_2) \end{vmatrix} \quad \begin{vmatrix} k_1 \\ k_2 \\ \frac{1}{3}(k_1 + 2k_2) \end{vmatrix}$$

Similarly for $\lambda = 5$,

$$\begin{vmatrix} -7 & 2 & -3 \\ 2 & -4 & -6 \\ -1 & 2 & -5 \end{vmatrix} \begin{vmatrix} x_1 \\ x_2 \\ x_3 \end{vmatrix} = \begin{vmatrix} 0 \\ 0 \\ 0 \end{vmatrix}$$

This can be reduced to

$$\begin{vmatrix} -1 & -2 & -5 \\ 0 & 16 & 32 \\ 0 & 0 & 0 \end{vmatrix} \begin{vmatrix} x_1 \\ x_2 \\ x_3 \end{vmatrix} = \begin{vmatrix} 0 \\ 0 \\ 0 \end{vmatrix}$$

$$-x_1 - 2x_2 - 5x_3 = 0$$
$$16x_2 + 32x_3 = 0 \tag{1.4}$$

There are two equations and three unknowns. Consider $x_3 = 1$. Then,

$$-x_1 - 2x_2 = 5$$
$$16x_2 = -32$$

Then, eigenvector for $\lambda = 5$ is

$$\begin{vmatrix} -1 \\ -2 \\ 1 \end{vmatrix}$$

Concept 1.1   *The eigenvectors are not unique. Equations (1.3) and (1.4) can be satisfied with other assumed numbers.*

1.3 DIAGONALIZATION OF A MATRIX

If a square matrix $\bar{A}$ of $n \times n$ has n linearly independent eigenvectors, then a matrix $\bar{P}$ can be found so that

$$\bar{P}^{-1}\bar{A}\bar{P} \tag{1.5}$$

is a diagonal matrix.

The matrix $\bar{P}$ is found by grouping the eigenvectors of $\bar{A}$ into a square matrix, that is, $\bar{P}$ has eigenvalues of $\bar{A}$ as its diagonal elements.

1.4 SIMILARITY TRANSFORMATION

The transformation of matrix $\bar{A}$ into $\bar{P}^{-1}A\bar{P}$ is called a *similarity transformation.* Diagonalization is a special case of similarity transformation.

Example 1.2   Continuing with Example 1.1, in Equation (1.4), by assuming that $x_3 = 1$, and solving, one eigenvector is

$$(-1, -2, 1)^t$$

Similarly in Equation (1.3), choose arbitrarily $x_1 = 2$, $x_2 = -1$, then

$$(2, -1, 0)^t$$

is an eigenvector. Similarly,

$$(3, 0, 1)^t$$

can be the third eigenvector. A matrix formed of these vectors is

$$\bar{P} = \begin{vmatrix} -1 & 2 & 3 \\ -2 & -1 & 0 \\ 1 & 0 & 1 \end{vmatrix}$$

and the diagonalization is obtained by application of Equation (1.5):

$$\bar{P}^{-1}\bar{A}\bar{P} = \begin{vmatrix} 5 & 0 & 0 \\ 0 & -3 & 0 \\ 0 & 0 & -3 \end{vmatrix}$$

This contains the eigenvalues as the diagonal elements.

Now choose some other eigenvectors and form a new matrix, say

$$\bar{P} = \begin{vmatrix} 1 & 1 & 3 \\ 2 & 1 & 0 \\ -1 & 1 & 1 \end{vmatrix}$$

Again with these values, $\bar{P}^{-1}A\bar{P}$ is

$$
\begin{vmatrix} 1 & 1 & 3 \\ 2 & 1 & 0 \\ -1 & 1 & 1 \end{vmatrix}^{-1}
\begin{vmatrix} -2 & 2 & -3 \\ 2 & 1 & -6 \\ -1 & -2 & 0 \end{vmatrix}
\begin{vmatrix} 1 & 1 & 3 \\ 2 & 1 & 0 \\ -1 & 1 & 1 \end{vmatrix} =
\begin{vmatrix} 5 & 0 & 0 \\ 0 & -3 & 0 \\ 0 & 0 & -3 \end{vmatrix}
$$

This is the same result as before.

Concept 1.2   *The sets of equations are an uncoupled system, which can be obtained with different eigenvectors.*

1.5 DECOUPLING A THREE-PHASE SYMMETRICAL SYSTEM

Let us decouple a three-phase transmission line section, where each phase has a mutual coupling with respect to ground. This is shown in Figure 1.1a. An impedance matrix of the three-phase transmission line can be written as

$$
\begin{vmatrix} Z_{aa} & Z_{ab} & Z_{ac} \\ Z_{ba} & Z_{bb} & Z_{bc} \\ Z_{ca} & Z_{cb} & Z_{cc} \end{vmatrix} \tag{1.6}
$$

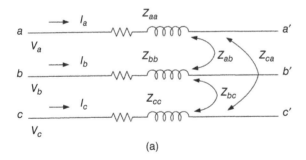

(a)

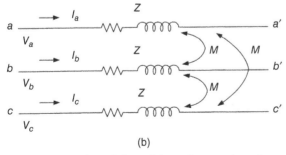

(b)

Figure 1.1   (a) An unbalanced three-phase section of a transmission line; (b) a balanced section.

where Z_{aa}, Z_{bb}, and Z_{cc} are the self-impedances of the phases a, b, and c; Z_{ab} is the mutual impedance between phases a and b, and Z_{ba} is the mutual impedance between phases b and a.

Assume that the line is *perfectly symmetrical*. This means all the mutual impedances, $Z_{ab} = Z_{ba} = M$, and all the self-impedances, $Z_{aa} = Z_{bb} = Z_{cc} = Z$, are equal (Figure 1.1b). This reduces the impedance matrix to

$$\begin{vmatrix} Z & M & M \\ M & Z & M \\ M & M & Z \end{vmatrix} \tag{1.7}$$

It is required to decouple this system using symmetrical components. First, find the eigenvalues:

$$\begin{vmatrix} Z - \lambda & M & M \\ M & Z - \lambda & M \\ M & M & Z - \lambda \end{vmatrix} = 0 \tag{1.8}$$

The eigenvalues are

$$\begin{aligned} \lambda &= Z + 2M \\ &= Z - M \\ &= Z - M \end{aligned} \tag{1.9}$$

The eigenvectors can be found by making $\lambda = Z + 2M$ and then $Z - M$. Substituting $\lambda = Z + 2M$,

$$\begin{vmatrix} Z - (Z + 2M) & M & M \\ M & Z - (Z + 2M) & M \\ M & M & Z - (Z + 2M) \end{vmatrix} \begin{vmatrix} X_1 \\ X_2 \\ X_3 \end{vmatrix} = 0 \tag{1.10}$$

By manipulation of the rows, this can be reduced to

$$\begin{vmatrix} -2 & 1 & 1 \\ 0 & -1 & 1 \\ 0 & 0 & 0 \end{vmatrix} \begin{vmatrix} X_1 \\ X_2 \\ X_3 \end{vmatrix} = 0 \tag{1.11}$$

This means that

$$\begin{aligned} -2X_1 + X_2 + X_3 &= 0 \\ -X_2 + X_3 &= 0 \\ X_3 &= 0 \end{aligned}$$

This gives $X_1 = X_2 = X_3 =$ any arbitrary constant k. Thus, one of the eigenvectors of the impedance matrix is

$$\begin{vmatrix} k \\ k \\ k \end{vmatrix} \tag{1.12}$$

It can be called the zero sequence eigenvector of the symmetrical component transformation matrix and can be written as

$$\begin{vmatrix} 1 \\ 1 \\ 1 \end{vmatrix} \tag{1.13}$$

Similarly for $\lambda = Z - M$,

$$\begin{vmatrix} Z-(Z-M) & M & M \\ M & Z-(Z-M) & M \\ M & M & Z-(Z-M) \end{vmatrix} \begin{vmatrix} X_1 \\ X_2 \\ X_3 \end{vmatrix} = 0 \tag{1.14}$$

which gives

$$\begin{vmatrix} 1 & 1 & 1 \\ 0 & 0 & 0 \\ 0 & 0 & 0 \end{vmatrix} \begin{vmatrix} X_1 \\ X_2 \\ X_3 \end{vmatrix} = 0 \tag{1.15}$$

$$X_1 + X_2 + X_3 = 0$$
$$0X_2 + 0X_3 = 0$$
$$0X_3 = 0$$

This gives the general relation $X_1 + X_2 + X_3 = 0$. Any choice of X_1, X_2, X_3, which satisfies this relation is a solution vector. Some choices are shown below:

$$\begin{vmatrix} X_1 \\ X_2 \\ X_3 \end{vmatrix} = \begin{vmatrix} 1 \\ a^2 \\ a \end{vmatrix}, \begin{vmatrix} 1 \\ a \\ a^2 \end{vmatrix}, \begin{vmatrix} 0 \\ \sqrt{3}/2 \\ -\sqrt{3}/2 \end{vmatrix}, \begin{vmatrix} 1 \\ -1/2 \\ -1/2 \end{vmatrix}, \text{ and so on,} \tag{1.16}$$

where a is a unit vector operator, which rotates by 120 degrees in the counterclockwise direction, that is,

$$a = -0.5 + j0.866$$
$$a^2 = -0.5 - j0.866$$
$$a^3 = 1 \tag{1.17}$$
$$1 + a + a^2 = 0$$

Concept 1.3 *Equation (1.16) is an important result and shows that, for perfectly symmetrical systems, multiple eigenvectors can be constructed for the transformation.*

1.6 SYMMETRICAL COMPONENT TRANSFORMATION

The voltage drop across a symmetrical section of a three-phase line can be written as

$$\begin{vmatrix} \Delta V_a \\ \Delta V_b \\ \Delta V_c \end{vmatrix} = \begin{vmatrix} Z & M & M \\ M & Z & M \\ M & M & Z \end{vmatrix} \begin{vmatrix} \Delta I_a \\ \Delta I_b \\ \Delta I_c \end{vmatrix} \tag{1.18}$$

Based on discussions in Section 1.5, the symmetrical component transformation eigenvectors are chosen as

$$
\begin{vmatrix} 1 \\ 1 \\ 1 \end{vmatrix} \begin{vmatrix} 1 \\ a \\ a^2 \end{vmatrix} \begin{vmatrix} 1 \\ a^2 \\ a \end{vmatrix}
\tag{1.19}
$$

A symmetrical component transformation matrix can, therefore, be written as

$$
\bar{T}_s = \begin{vmatrix} 1 & 1 & 1 \\ 1 & a^2 & a \\ 1 & a & a^2 \end{vmatrix}
\tag{1.20}
$$

Its inverse is

$$
\bar{T}_s^{-1} = \frac{1}{3} \begin{vmatrix} 1 & 1 & 1 \\ 1 & a & a^2 \\ 1 & a^2 & a \end{vmatrix}
\tag{1.21}
$$

For the transformation of currents, we can writ

$$
\begin{vmatrix} I_a \\ I_b \\ I_c \end{vmatrix} = \begin{vmatrix} 1 & 1 & 1 \\ 1 & a^2 & a \\ 1 & a & a^2 \end{vmatrix} \begin{vmatrix} I_0 \\ I_1 \\ I_2 \end{vmatrix}
\tag{1.22}
$$

Or in abbreviated form, we can write

$$
\bar{I}_{abc} = \bar{T}_s \bar{I}_{012}
\tag{1.23}
$$

where $\bar{I}_{abc}$ are the original currents in phases a, b, and c, which are transformed into zero sequence, positive sequence, and negative sequence currents, $\bar{I}_{012}$. The original phasors are subscripted abc and the sequence components are subscripted 012. Similarly, for transformation of voltages,

$$
\begin{vmatrix} \Delta V_a \\ \Delta V_b \\ \Delta V_c \end{vmatrix} = \begin{vmatrix} 1 & 1 & 1 \\ 1 & a^2 & a \\ 1 & a & a^2 \end{vmatrix} \begin{vmatrix} \Delta V_0 \\ \Delta V_1 \\ \Delta V_2 \end{vmatrix}
\tag{1.24}
$$

Or in the abbreviated form

$$
\Delta \bar{V}_{abc} = \bar{T}_s \Delta \bar{V}_{012}
\tag{1.25}
$$

Or in general

$$
\bar{V}_{abc} = \bar{T}_s \bar{V}_{012}
\tag{1.26}
$$

where $\bar{V}_{abc}$ are the original voltages in phases a, b, and c, which are transformed into zero sequence, positive sequence, and negative sequence voltages, $\bar{V}_{012}$. The original phasors are subscripted abc and the sequence components are subscripted 012.

Conversely,

$$
\bar{I}_{012} = \bar{T}_s^{-1} \bar{I}_{abc}, \qquad \bar{V}_{012} = \bar{T}_s^{-1} \bar{V}_{abc}
\tag{1.27}
$$

The transformation of impedance is not straightforward and is derived as follows:

$$\bar{V}_{abc} = \bar{Z}_{abc}\bar{I}_{abc}$$
$$\bar{T}_s\bar{V}_{012} = \bar{Z}_{abc}\bar{T}_s\bar{I}_{012}$$
$$\bar{V}_{012} = \bar{T}_s^{-1}\bar{Z}_{abc}\bar{T}_s\bar{I}_{012} = \bar{Z}_{012}\bar{I}_{012}$$

(1.28)

Therefore,

$$\bar{Z}_{012} = \bar{T}_s^{-1}\bar{Z}_{abc}\bar{T}_s$$

(1.29)

$$\bar{Z}_{abc} = \bar{T}_s\bar{Z}_{012}\bar{T}_s^{-1}$$

(1.30)

Applying the impedance transformation to the original impedance matrix of the three-phase symmetrical transmission line in Equation (1.7), the transformed matrix is

$$
\bar{Z}_{012} = \frac{1}{3}
\begin{vmatrix}
1 & 1 & 1 \\
1 & a & a^2 \\
1 & a^2 & a
\end{vmatrix}
\begin{vmatrix}
Z & M & M \\
M & Z & M \\
M & M & Z
\end{vmatrix}
\begin{vmatrix}
1 & 1 & 1 \\
1 & a^2 & a \\
1 & a & a^2
\end{vmatrix}
$$
$$
=
\begin{vmatrix}
Z + 2M & 0 & 0 \\
0 & Z - M & 0 \\
0 & 0 & Z - M
\end{vmatrix}
$$

(1.31)

The original three-phase coupled system has been decoupled through symmetrical component transformation. It is diagonal, and all off-diagonal terms are zero, meaning that there is no coupling between the sequence components. Decoupled positive, negative, and zero sequence networks are shown in Figure 1.2. The transformations described above are called symmetrical component transformations.

1.7 DECOUPLING A THREE-PHASE UNSYMMETRICAL SYSTEM

Now consider that the original three-phase system is not completely balanced. *Ignoring the mutual impedances* in Equation (1.6), let us assume unequal phase impedances, Z_1, Z_2, and Z_3, that is, the impedance matrix is

$$
\bar{Z}_{abc} =
\begin{vmatrix}
Z_1 & 0 & 0 \\
0 & Z_2 & 0 \\
0 & 0 & Z_3
\end{vmatrix}
$$

(1.32)

The symmetrical component transformation is

$$
\bar{Z}_{012} = \frac{1}{3}
\begin{vmatrix}
1 & 1 & 1 \\
1 & a & a^2 \\
1 & a^2 & a
\end{vmatrix}
\begin{vmatrix}
Z_1 & 0 & 0 \\
0 & Z_2 & 0 \\
0 & 0 & Z_3
\end{vmatrix}
\begin{vmatrix}
1 & 1 & 1 \\
1 & a^2 & a \\
1 & a & a^2
\end{vmatrix}
$$
$$
= \frac{1}{3}
\begin{vmatrix}
Z_1 + Z_2 + Z_3 & Z_1 + a^2 Z_2 + a Z_3 & Z_1 + a Z_2 + Z_3 \\
Z_1 + a Z_2 + a Z_3 & Z_1 + Z_2 + Z_3 & Z_1 + a^2 Z_2 + a Z_3 \\
Z_1 + a^2 Z_2 + a Z_3 & Z_1 + a Z_2 + a Z_3 & Z_1 + Z_2 + Z_3
\end{vmatrix}
$$

(1.33)

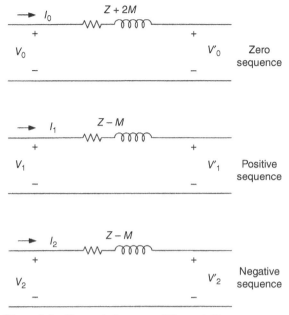

Figure 1.2   Decoupled system of Figure 1.1b.

The resulting matrix shows that the *original unbalanced system is not decoupled.* If we start with equal self-impedances and unequal mutual impedances or vice versa, the resulting matrix is nonsymmetrical. It is a minor problem today, as nonreciprocal networks can be easily handled on digital computers. Nevertheless, the main application of symmetrical components is for the study of unsymmetrical faults. Negative sequence relaying, stability calculations, and machine modeling are some other examples.

Concept 1.4   *It is assumed that the system is perfectly symmetrical before an unbalance condition occurs. The asymmetry occurs only at the fault point. The symmetrical portion of the network is considered to be isolated, to which an unbalanced condition is applied at the fault point. In other words, the unbalance part of the network can be thought to be connected to the balanced system at the point of fault. Practically, the power systems are not perfectly balanced and some asymmetry always exists. However, the error introduced by ignoring this asymmetry is, generally, small.*

1.8 CLARKE COMPONENT TRANSFORMATION

Review the concept outlined in Section 1.5, *that eigenvectors are not unique.* The Clarke component transformation is defined as

$$\begin{vmatrix} V_a \\ V_b \\ V_c \end{vmatrix} = \begin{vmatrix} 1 & 1 & 0 \\ 1 & -\dfrac{1}{2} & \dfrac{\sqrt{3}}{2} \\ 1 & -\dfrac{1}{2} & -\dfrac{\sqrt{3}}{2} \end{vmatrix} \begin{vmatrix} V_0 \\ V_\alpha \\ V_\beta \end{vmatrix} \tag{1.34}$$

Note that the eigenvalues satisfy the relations derived in Equation (1.16), and

$$
\begin{vmatrix} V_0 \\ V_\alpha \\ V_\beta \end{vmatrix} = \begin{vmatrix} \dfrac{1}{3} & \dfrac{1}{3} & \dfrac{1}{3} \\ \dfrac{2}{3} & -\dfrac{1}{3} & -\dfrac{1}{3} \\ 0 & \dfrac{1}{\sqrt{3}} & -\dfrac{1}{\sqrt{3}} \end{vmatrix} \begin{vmatrix} V_a \\ V_b \\ V_c \end{vmatrix} \tag{1.35}
$$

Similar equations may be written for the current. Note that

$$
\begin{vmatrix} 1 \\ 1 \\ 1 \end{vmatrix}, \quad \begin{vmatrix} 1 \\ -\dfrac{1}{2} \\ -\dfrac{1}{2} \end{vmatrix}, \quad \begin{vmatrix} 0 \\ \dfrac{\sqrt{3}}{2} \\ -\dfrac{\sqrt{3}}{2} \end{vmatrix} \tag{1.36}
$$

are the eigenvectors of a perfectly symmetrical impedance.

The transformation matrices are

$$
\bar{T}_c = \begin{vmatrix} 1 & 1 & 0 \\ 1 & -1/2 & \sqrt{3}/2 \\ 1 & 1/2 & \sqrt{3}/2 \end{vmatrix} \tag{1.37}
$$

$$
\bar{T}_c^{-1} = \begin{vmatrix} 1/3 & 1/3 & 1/3 \\ 2/3 & -1/3 & -1/3 \\ 0 & 1/\sqrt{3} & -1/\sqrt{3} \end{vmatrix} \tag{1.38}
$$

and as before,

$$
\bar{Z}_{0\alpha\beta} = \bar{T}_c^{-1} \bar{Z}_{abc} \bar{T}_c \tag{1.39}
$$
$$
\bar{Z}_{abc} = \bar{T}_c \bar{Z}_{0\alpha\beta} \bar{T}_c^{-1} \tag{1.40}
$$

The Clarke component expression for a perfectly symmetrical system is

$$
\begin{vmatrix} V_0 \\ V_\alpha \\ V_\beta \end{vmatrix} = \begin{vmatrix} Z_{00} & 0 & 0 \\ 0 & Z_{\alpha\alpha} & 0 \\ 0 & 0 & Z_{\beta\beta} \end{vmatrix} \begin{vmatrix} I_0 \\ I_\alpha \\ I_\beta \end{vmatrix} \tag{1.41}
$$

The same philosophy of transformation can also be applied to systems with two or more three-phase circuits in parallel. The instantaneous power theory and EMTP (Electromagnetic Transient Program) modeling of transmission lines, are based upon Clarke's transformation [9, 10].

1.9 SIGNIFICANCE OF SELECTION OF EIGENVECTORS IN SYMMETRICAL COMPONENTS

The significance of symmetrical components will be illustrated with an example.

Example 1.3   As the vector selection is arbitrary, design an entirely new transformation system and apply it to a transformation.

Let the eigenvectors be

$$\begin{vmatrix} 1 \\ 1 \\ 1 \end{vmatrix}, \begin{vmatrix} 1 \\ -1/4 \\ -3/4 \end{vmatrix}, \begin{vmatrix} 1 \\ -1/2 \\ -1/2 \end{vmatrix}$$

Then the transformation matrix is

$$\bar{T}_t = \begin{vmatrix} 1 & 1 & 1 \\ 1 & -1/4 & -1/2 \\ 1 & -3/4 & -1/2 \end{vmatrix} \tag{1.42}$$

Its inverse is

$$\bar{T}_t^{-1} = \begin{vmatrix} 1/3 & 1/3 & 1/3 \\ 0 & 2 & -2 \\ 2/3 & -2\frac{1}{3} & 1\frac{2}{3} \end{vmatrix} \tag{1.43}$$

Then,

$$\bar{Z}_{012} = \begin{vmatrix} 1/3 & 1/3 & 1/3 \\ 0 & 2 & -2 \\ 2/3 & -2\frac{1}{3} & 1\frac{2}{3} \end{vmatrix} \begin{vmatrix} Z & M & M \\ M & Z & M \\ M & M & Z \end{vmatrix} \begin{vmatrix} 1 & 1 & 1 \\ 1 & -0.25 & -0.5 \\ 1 & -0.75 & -0.5 \end{vmatrix}$$

$$= \begin{vmatrix} Z+2M & 0 & 0 \\ 0 & Z-M & 0 \\ 0 & 0 & Z-M \end{vmatrix} \tag{1.44}$$

This gives the same impedance transformation as with the symmetrical component vectors. Rather than being thrilled with this result let us examine further.

Let us denote the voltages in a balanced three-phase system as

$$\begin{aligned} V_a &= V < 0° \\ V_b &= V < -120° \\ V_c &= V < 120° \end{aligned} \tag{1.45}$$

This is the standard representation in a three-phase system with counterclockwise rotation of the vectors. The "rotation" signifies the direction in which the vectors rotate. The "sequence" describes the order in which the three-phase vectors rotate, that is, it can be ABC, or ACB, see Chapter 2. Then, first apply symmetrical component transformation:

$$\begin{vmatrix} V_0 \\ V_1 \\ V_2 \end{vmatrix} = \frac{1}{3} \begin{vmatrix} 1 & 1 & 1 \\ 1 & -0.5+j0.866 & -0.5-j0.866 \\ 1 & -0.5-j0.866 & -0.5+j0.866 \end{vmatrix} \begin{vmatrix} V \\ V(-0.5-j0.866) \\ V(-0.5+j0.866) \end{vmatrix} = \begin{vmatrix} 0 \\ V \\ 0 \end{vmatrix} \tag{1.46}$$

That is, zero and negative sequence components are zero. In the symmetrical components, the operator "a" rotates a unity vector by 120 degrees electrical in the counterclockwise direction. By international convention, this is the accepted rotation of the vectors in a three-phase system. Also in a balanced system, each current and voltage vector is displaced by 120 electrical degrees. Thus, for a balanced system these vectors can be called as positive sequence vectors (components), and the *negative sequence and zero sequence components should be zero*, as illustrated in Equation (1.46).

Now apply the transformation in Equation (1.43)

$$
\begin{vmatrix} V_0 \\ V_1 \\ V_2 \end{vmatrix} = \begin{vmatrix} 1/3 & 1/3 & 1/3 \\ 0 & 2 & -2 \\ 2/3 & -2\frac{1}{3} & 1\frac{2}{3} \end{vmatrix} \begin{vmatrix} V \\ V(-0.5 - j0.866) \\ V(-0.5 + j0.866) \end{vmatrix} = \begin{vmatrix} 0 \\ -j3.5V \\ V(1 + j3.5) \end{vmatrix} \quad (1.47)
$$

Meaningful interpretation cannot be applied to the end result of Equation (1.47). It flouts the accepted convention and phasors in a three-phase system. We question the results obtained in this equation, as a three-phase unbalanced system does not generate any negative or zero sequence components. For further simplified explanation, see Chapter 2.

Concept 1.5 *In symmetrical component transformation, the eigenvectors are selected to provide meaningful physical interpretation. The scientific and engineering logic of symmetrical components lies in selection of appropriate eigenvectors.*

REFERENCES

[1] CL Fortescu. Method of symmetrical coordinates applied to the solution of polyphase networks. *Transactions of the American Institute of Electrical Engineers*, vol. 37, no. 2, pp. 1027–1140, July 1918.

[2] JW Nilsson and S Riedel. *Electrical Circuits*, 10th edition. Prentice Hall, 2014.

[3] CF Wagner and RD Evans. *Symmetrical Components*. New York: McGraw-Hill, 1933.

[4] E Clarke. *Circuit Analysis of Alternating Current Power Systems*, vol. 1. New York: Wiley, 1943.

[5] LJ Myatt. *Symmetrical Components*. Oxford, UK: Pergamon Press, 1968.

[6] JL Blackburn. *Symmetrical Components for Power System Engineering*. New York: Marcel and Dekker, 1993.

[7] WE Lewis and DG Pryce. *The Application of Matrix Theory to Electrical Engineering*. London: E&FN Spon, 1965, chapter 6.

[8] FR Gantmatcher. *Application of Matrix Theory of Matrices*. Dover, 2005. Originally published by Interscience Publishers, 1959.

[9] H Akagi and A Nabe. The *p-q* theory in three-phase systems under non-sinusoidal conditions. *European Transactions on Electrical Power*, vol. 3, no. 1, pp. 27–30, 1993.

[10] *ATP Rule Book*. Portland, OR: ATP User Group, 1992.

CHAPTER 2

FUNDAMENTAL CONCEPTS OF SYMMETRICAL COMPONENTS

IN **CHAPTER 1**, we resorted to matrix algebra to convey the basic concept that the symmetrical component transformation is not unique, because eigenvectors can be chosen arbitrarily. Section 1.9 enlarges this concept and notes that the selection of eigenvectors in the symmetrical component transformation has meaningful concepts of decoupling a three-phase unbalanced system.

Another way of looking at the basic theory of symmetrical components can be explained as a mathematical concept. A system of three coplanar vectors is completely defined by six parameters, and the system can be said to possess six degrees of freedom. A point in a straight line, being constrained to lie on the line, possesses only one degree of freedom, and by the same analogy, a point in space has three degrees of freedom. A coplanar vector is defined by its terminal and length and therefore possesses two degrees of freedom. A system of coplanar vectors having six degrees of freedom, that is, a three-phase unbalanced current or voltage vectors, can be represented by three symmetrical systems of vectors each having two degrees of freedom. In general, a system of n numbers can be resolved into n sets of component numbers each having n components, that is, a total of n^2 components. Fortescue demonstrated that an unbalanced set of n phasors can be resolved into $n - 1$ balanced phase systems of different phase sequence and one zero sequence system, in which all phasors are of equal magnitude and cophasial:

$$\begin{aligned}
V_a &= V_{a1} + V_{a2} + V_{a3} + \ldots + V_{an} \\
V_b &= V_{b1} + V_{b2} + V_{b3} + \ldots + V_{bn} \\
V_n &= V_{n1} + V_{n2} + V_{n3} + \ldots + V_{nn}
\end{aligned} \tag{2.1}$$

where V_a, V_b,..., V_n are original n unbalanced voltage phasors. V_{a1}, V_{b1},..., V_{n1} are the first set of n balanced phasors at an angle of $2\pi/n$ between them; V_{a2}, V_{b2},..., V_{n2} are the second set of n balanced phasors at an angle $4\pi/n$; and the final set V_{an}, V_{bn},..., V_{nn} is the zero sequence set, all phasors at $n(2\pi/n) = 2\pi$, that is, cophasial.

In a symmetrical three-phase balanced system, the generators produce balanced voltages which are displaced from each other by $2\pi/3 = 120°$. These voltages can be called positive sequence voltages. If a vector operator a is defined which rotates a unit vector through $120°$ in a counterclockwise direction, then $a = -0.5 + j0.866$,

Understanding Symmetrical Components for Power System Modeling, First Edition. J.C. Das.
© 2017 by The Institute of Electrical and Electronics Engineers, Inc. Published 2017 by John Wiley & Sons, Inc.

$a^2 = -0.5 - j0.866$, $a^3 = 1$, $1 + a^2 + a = 0$. Considering a three-phase system, Equation (2.1) reduce to

$$V_a = V_{a0} + V_{a1} + V_{a2}$$
$$V_b = V_{b0} + V_{b1} + V_{b2} \quad (2.2)$$
$$V_c = V_{c0} + V_{c1} + V_{c2}$$

We can define the set consisting of V_{a0}, V_{b0}, and V_{c0} as the zero sequence set, the set V_{a1}, V_{b1}, and V_{c1} as the positive sequence set, and the set V_{a2}, V_{b2}, and V_{c2} as the negative sequence set of voltages. The three original unbalanced voltage vectors give rise to nine voltage vectors, which must have constraints of freedom and are not totally independent. By definition of positive sequence, V_{a1}, V_{b1}, and V_{c1} should be related as follows, as in a normal balanced system:

$$V_{b1} = a^2 V_{a1}, V_{c1} = a V_{a1} \quad (2.3)$$

Note that V_{a1} phasor is taken as the reference vector.

The negative sequence set can be similarly defined, but of opposite phase sequence:

$$V_{b2} = a V_{a2}, V_{c2} = a^2 V_{a2} \quad (2.4)$$

Also, $V_{a0} = V_{b0} = V_{c0}$. With these relations defined, Equation (2.2) can be written as

$$\begin{vmatrix} V_a \\ V_b \\ V_c \end{vmatrix} = \begin{vmatrix} 1 & 1 & 1 \\ 1 & a^2 & a \\ 1 & a & a^2 \end{vmatrix} \begin{vmatrix} V_{a0} \\ V_{a1} \\ V_{a2} \end{vmatrix} \quad (2.5)$$

or in the abbreviated form:

$$\bar{V}_{abc} = \bar{T}_s \bar{V}_{012} \quad (2.6)$$

Conversely,

$$\bar{V}_{012} = \bar{T}_s^{-1} \bar{V}_{abc} \quad (2.7)$$

This is the same result as we obtained in Equations (1.23) and (1.26).

2.1 CHARACTERISTICS OF SYMMETRICAL COMPONENTS

Matrix equations (2.5) are written in the expanded form:

$$V_a = V_0 + V_1 + V_2$$
$$V_b = V_0 + a^2 V_1 + a V_2 \quad (2.8)$$
$$V_c = V_0 + a V_1 + a^2 V_2$$

and

$$V_0 = \frac{1}{3}(V_a + V_b + V_c)$$

$$V_1 = \frac{1}{3}(V_a + aV_b + a^2V_c) \tag{2.9}$$

$$V_2 = \frac{1}{3}(V_a + a^2V_b + aV_c)$$

It implies that three-phase *unbalanced* voltages V_a, V_b, and V_c can be resolved into three voltages, V_0, V_1, and V_2, defined as follows:

- V_0 is the zero sequence voltage. It is of equal magnitude in all the three phases and is cophasial.
- V_1 is the system of balanced positive sequence voltages, *of the same phase sequence* as the original unbalanced system of voltages. It is of equal magnitude in each phase, but displaced by 120°, the component of phase b lagging the component of phase a by 120°, and the component of phase c leading the component of phase a by 120°.
- V_2 is the system of balanced negative sequence voltages. It is of equal magnitude in each phase, and there is a 120° phase displacement between the voltages, the component of phase c lagging the component of phase a, and the component of phase b leading the component of phase a.

Therefore, the positive and negative sequence voltages (or currents) can be defined as "the order in which the three phases attain a maximum value." For the positive sequence the order is *abca* while for the negative sequence it is *acba*. We can also define positive and negative sequence by the order in which the phasors pass a *fixed point, a stationary observer in space.*

Concept 2.1   *Note that the rotation is counterclockwise for all three sets of sequence components, as was assumed for the original unbalanced vectors. Sometimes, this is confused and negative sequence rotation is said to be the reverse of positive sequence. The negative sequence vectors do not rotate in a direction opposite to the positive sequence vectors, though the negative phase sequence is opposite to the positive phase sequence.*

Example 2.1   Consider a system of unbalanced currents given by

$$I_a = 0.8 < 10°$$
$$I_b = 1.25 < -70°$$
$$I_c = 0.9 < 130°$$

The original unbalanced current phasors are shown in Figure 2.1a, and the resolution of these unbalanced currents into symmetrical components is shown in Figure 2.1b. A reader can prove using the above equations that then the transformation back from the symmetrical components to original phasors will give results as shown in Figure 2.1a.

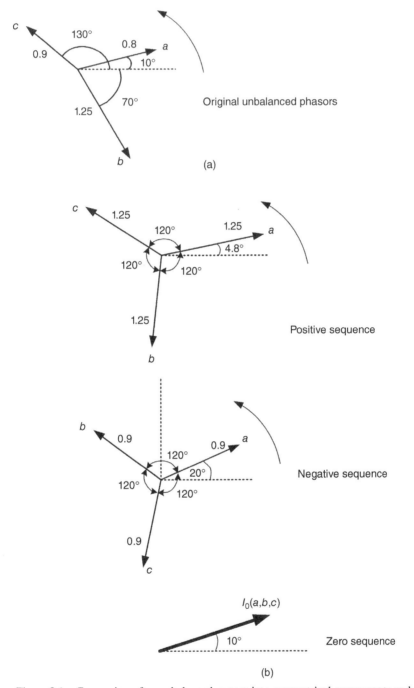

Figure 2.1   Conversion of an unbalanced system into symmetrical components and from symmetrical components back to original unbalanced system. (a) Original unbalanced system; (b) conversion to symmetrical components.

Concept 2.2   *It may seem more complex to resolve the original three-phase unbal-anced currents or voltages consisting of three phasors to three symmetrical compo-nent subsystems consisting of total of nine phasors. However, note that these nine phasors are related to each other, in fact, acting as three phasors. It is recognized that it is far easier to analyze three balanced systems and then revert back to original system than to directly analyze a three-phase unbalanced system. This will be further documented in this book.*

2.2 CHARACTERISTICS OF SEQUENCE NETWORKS

1. In a three-phase unfaulted system in which all loads are balanced and in which generators produce positive sequence voltages, only positive sequence currents flow, resulting in balanced voltage drops of the same sequence. There are no negative sequences or zero sequence voltage drops.

2. The currents and voltages of different sequences do not affect each other, that is, *positive sequence currents produce only positive sequence voltage drops. By the same analogy, the negative sequence currents produce only neg-ative sequence drops, and zero sequence currents produce only zero sequence drops.*

3. Negative and zero sequence currents are set up in circuits of unbalanced impedances only, that is, a set of unbalanced impedances in a symmetrical sys-tem may be regarded as a source of negative and zero sequence currents. Posi-tive sequence currents flowing in an unbalanced system produce positive, nega-tive, and possibly zero sequence voltage drops. The negative sequence currents flowing in an unbalanced system produce voltage drops of all three sequences. The same is true about zero sequence currents.

4. In a three-phase three-wire system, no zero sequence currents appear in the line conductors. This is so because $I_0 = (1/3)(I_a + I_b + I_c)$ and, therefore, there is no path for the zero sequence current to flow. In a three-phase, four-wire system with neutral return, the neutral must carry out-of-balance current, that is, $I_n = (I_a + I_b + I_c)$. Therefore, it follows that $I_n = 3I_0$. At the grounded neutral of a three-phase wye system, positive and negative sequence voltages are zero. The neutral voltage is equal to the zero sequence voltage or product of zero sequence current and three times the neutral impedance Z_n.

5. From what has been said in point 4 above, phase conductors emanat-ing from ungrounded wye or delta connected transformer windings can-not have zero sequence current. In a delta winding, zero sequence currents, if present, set up circulating currents in the delta winding itself. This is because the delta winding forms a closed path of low impedance for the zero sequence currents; each phase zero sequence voltage is absorbed by its own phase voltage drop and there are no zero sequence components at the terminals.

2.3 SEQUENCE IMPEDANCE OF NETWORK COMPONENTS

The impedance encountered by the symmetrical components depends on the type of power system equipment, that is, a generator, a transformer, or a transmission line. The sequence impedances are required for component modeling and analysis. We derived the sequence impedances of a symmetrical coupled transmission line in Equation (1.31). Zero sequence impedance of overhead lines depends on the presence of ground wires, tower footing resistance, and grounding. It may vary between two and six times the positive sequence impedance. The line capacitance of overhead lines is ignored in short-circuit calculations. Chapter 3 details three-phase matrix models of transmission lines, bundle conductors, and cables, and their transformation into symmetrical components. While estimating sequence impedances of power system components is one problem, constructing the zero, positive, and negative sequence impedance networks is the first step for unsymmetrical fault current calculations.

2.4 CONSTRUCTION OF SEQUENCE NETWORKS

Concept 2.3 *Positive sequence currents only flow in positive sequence network and not in negative or zero sequence networks. Negative sequence currents flow only in negative sequence network and not in positive or zero sequence networks. Zero sequence currents flow only in zero sequence networks and not in positive or negative sequence networks. Therefore it becomes necessary to construct positive, negative, and zero sequence networks.*

A sequence network shows how the sequence currents, if these are present, will flow in a system. Connections between sequence component networks are necessary to achieve this objective. The sequence networks are constructed as viewed from the *fault point,* which can be defined as the point at which the unbalance occurs in a system, that is, a fault or load unbalance.

The voltages for the sequence networks are taken as line-to-neutral voltages. The only active network containing the voltage source is the positive sequence network. Phase *a* voltage is taken as the reference voltage, and the voltages of the other two phases are expressed with reference to phase *a* voltage.

The sequence networks for positive, negative, and zero sequence will have per phase impedance values which may differ. Normally, the sequence impedance networks are constructed on the basis of per unit values on a common MVA base, and a base MVA of 100 is in common use. For non-rotating equipment like transformers, the impedance to negative sequence currents will be the same as for positive sequence currents. The impedance to negative sequence currents of rotating equipment will be different from the positive sequence impedance and, in general, for all apparatuses the impedance to zero sequence currents will be different from the positive or

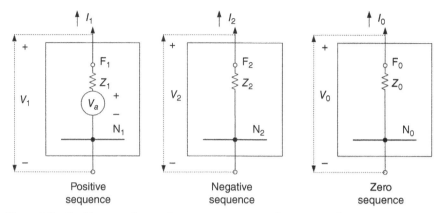

Figure 2.2   Positive, negative, and zero sequence network representation.

negative sequence impedances. For a study involving sequence components, the sequence impedance data can be

1. calculated by using subroutine computer programs,
2. obtained from manufacturers' data,
3. calculated by long-hand calculations, or
4. estimated from tables in published references.

The positive direction of current flow in each sequence network is *outward* at the faulted or unbalance point. This means that the sequence currents flow in the same direction in all the three sequence networks.

Sequence networks are shown schematically in boxes in which the fault points from which the sequence currents flow outward are marked as F_1, F_2, and F_0, and the neutral buses are designated as N_1, N_2, and N_0, respectively, for the positive, negative, and zero sequence impedance networks. Each network forms a two-port network with Thévenin sequence voltages across sequence impedances. Figure 2.2 illustrates this basic formation. Note the direction of currents. The voltage across the sequence impedance rises from N to F. As stated before, only the positive sequence network has a voltage source, which is the Thévenin equivalent. With this convention, appropriate signs must be allocated to the sequence voltages

$$V_1 = V_a - I_1 Z_1$$
$$V_2 = -I_2 Z_2$$
$$V_0 = -I_0 Z_0$$

(2.10)

or in matrix form

$$\begin{vmatrix} V_0 \\ V_1 \\ V_2 \end{vmatrix} = \begin{vmatrix} 0 \\ V_a \\ 0 \end{vmatrix} - \begin{vmatrix} Z_0 & 0 & 0 \\ 0 & Z_1 & 0 \\ 0 & 0 & Z_2 \end{vmatrix} \begin{vmatrix} I_0 \\ I_1 \\ I_2 \end{vmatrix}$$

(2.11)

Based on the discussions so far, we can graphically represent the sequence impedances of various system components. Also see References [1–7].

Concept 2.4 *What has been said above implies that three impedance networks —* *positive, negative, and zero sequence networks — must be constructed. That means* *the values for each of the system components should be calculated. While positive and* *negative sequence networks are continuous, (means have no discontinuity) the zero* *sequence network is not. The transformer zero sequence impedance can introduce* *discontinuities.*

2.5 SEQUENCE COMPONENTS OF TRANSFORMERS

We will discuss the sequence impedances of the various power system components in this book. Here, we will address the sequence impedances of two-winding and three-winding transformers.

The positive and negative sequence impedances of a transformer can be taken to be equal to its leakage impedance. As the transformer is a static device, the positive or negative sequence impedances do not change with phase sequence of the applied balanced voltages. The zero sequence impedance can, however, vary from an open circuit to a low value depending on the transformer winding connection, method of neutral grounding, and transformer construction, that is, core or shell type.

We will briefly discuss the shell and core form of construction, as it has a major impact on the zero sequence flux and impedance. Referring to Figure 2.3a, in a three-phase core-type transformer, the sum of the fluxes in each phase in a given direction along the cores is zero; however, the flux going up one leg of the core must return through the other two, that is, the magnetic circuit of a phase is completed through the other two phases in parallel. The magnetizing current per phase is that required for the core and part of the yoke. This means that in a three-phase core-type transformer, the magnetizing current will be different in each phase. Generally, the cores are long compared to yokes and the yokes are of greater cross section. The yoke reluctance is only a small fraction of the core and the variation of magnetizing current per phase is not appreciable. However, consider now the zero sequence flux, which will be directed in one direction, in each of the core legs. The return path lies, not through the core legs, but through insulating medium and tank.

In three separate single-phase transformers connected in three-phase configuration or in shell-type three-phase transformers, the magnetic circuits of each phase are complete in themselves and do not interact(Figure 2.3b). Due to advantages in short-circuit and transient voltage performance, the shell form is used for larger transformers. The variations in shell form have five- or seven-legged cores. *Briefly, we can* *say that, in a core type, the windings surround the core, and in the shell type, the core* *surrounds the windings.*

2.5.1 Delta-Wye or Wye-Delta Transformer

In a delta-wye transformer with the wye winding grounded, zero sequence impedance is taken approximately equal to positive or negative sequence impedance, viewed

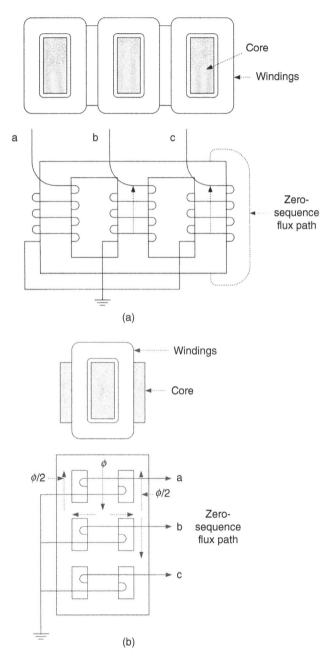

Figure 2.3   (a) Core form of three-phase transformers, flux paths for phase, and zero sequence currents; (b) shell form of three-phase transformers.

from the wye connection side. In fact, the impedance to the flow of zero sequence currents in the core-type transformers is lower as compared to the positive sequence impedance. This is so, because there is no return path for zero sequence exciting flux in core-type units except through insulating medium and tank, a path of high reluctance. In groups of three single-phase transformers or in three-phase shell-type transformers, the zero sequence impedance is higher.

The zero sequence networks for a wye-delta transformer are shown in Figure 2.4a. The grounding of the wye neutral allows the zero sequence currents to return through the neutral and circulate in the windings to the source of unbalance. Thus, the circuit on the wye side is shown connected to the L side line. On the delta side, the circuit is open, as no zero sequence currents appear in the lines, though these currents circulate in the delta windings to balance the ampere turns in the wye windings. The circuit is open on the H side line, and the zero sequence impedance of the transformer seen from the high side is an open circuit. If the wye winding neutral is left isolated (ungrounded), Figure 2.4b, the circuit will be open on both sides, presenting infinite impedance.

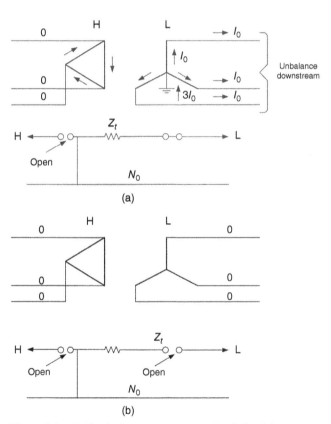

Figure 2.4   (a) Equivalent zero sequence circuit for delta-wye transformer, neutral solidly grounded; (b) zero sequence circuit of delta-wye transformer, wye neutral isolated.

Three-phase current flow diagrams can be constructed based on the convention that current always flows to the unbalance and that the ampere turns in primary windings must be balanced by the ampere turns in the secondary windings.

2.5.2 Wye-Wye Transformer

In a wye-wye connected transformer, with both neutrals isolated, no zero sequence currents can flow. The zero sequence equivalent circuit is open on *both* sides and presents infinite impedance to the flow of zero sequence currents. When one of the neutrals is grounded, still no zero sequence currents can be transferred from the grounded side to the ungrounded side. With one neutral grounded, there are no balancing ampere turns in the ungrounded wye windings to enable current to flow in the grounded neutral windings. Thus, neither of the windings can carry a zero sequence current. Both neutrals must be grounded for the transfer of zero sequence currents.

A wye-wye connected transformer with isolated neutrals is not used due to the phenomenon of the oscillating neutral (Figure 2.5). Due to saturation in transformers and the flat-topped flux wave, a peak EMF is generated which does not balance the applied sinusoidal voltage and generates a resultant third (and other) harmonics. These distort the transformer voltages as the neutral oscillates at thrice the supply frequency, a phenomenon called the "oscillating neutral." A tertiary delta is added to circulate the third harmonic currents and stabilize the neutral. It may also be designed as a load winding, which may have a rated voltage distinct from high- and low-voltage windings. When provided for zero sequence current circulation and harmonic suppression, the terminals of the tertiary connected delta winding may not be brought out of the transformer tank. Sometimes core-type transformers are provided with five-legged cores to circulate the harmonic currents.

2.5.3 Delta-Delta Transformer

In a delta-delta connection, no zero currents will pass from one winding to another. On the transformer side, the windings are shown connected to the reference bus, allowing the circulation of currents within the windings on the line side connections are open.

2.5.4 Zigzag Transformer

A zigzag transformer is often used to derive a neutral for grounding of a delta-delta connected system. This is shown in Figure 2.6. Windings a_1 and a_2 are on the same

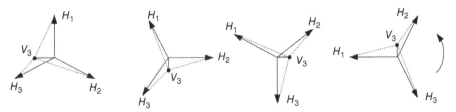

Figure 2.5   Phenomenon of oscillating neutral in wye-wye connected transformer, both neutrals isolated.

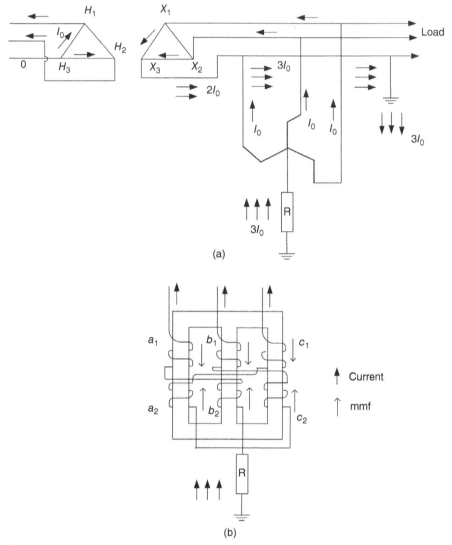

Figure 2.6   Current distribution in a delta-delta system with zigzag grounding transformer for a single line-to-ground fault; (b) zigzag transformer winding connections.

core leg and have the same number of turns but are wound in the opposite direction. The zero sequence currents in the two windings on the same core leg have canceling ampere turns. Referring to Figure 2.6b, the currents in the winding sections a_1 and c_2 must be equal as these are in series. By the same analogy all currents must be equal, balancing the MMF in each leg:

$$i_{a1} = i_{a2} = i_{b1} = i_{b2} = i_{c1} = i_{c2} \tag{2.12}$$

The impedance to the zero sequence currents is that due to leakage flux of the windings. For positive or negative sequence currents, neglecting magnetizing current, the connection has infinite impedance. Figure 2.6a shows the distribution of zero sequence current and its return path for a single line-to-ground fault on one of the phases. The ground current divides equally through the zigzag transformer; one-third of the current returns directly to the fault point and the remaining two-thirds must pass through two phases of the delta connected windings to return to the fault point. Two phases and windings on the primary delta must carry current to balance the ampere turns of the secondary-winding currents (Figure 2.6b). An impedance can be added between the artificially derived neutral and ground to limit the ground fault current.

Table 2.1 shows the sequence equivalent circuits of three-phase two-winding transformers. When the transformer neutral is grounded through an impedance Z_n, a term $3Z_n$ appears in the equivalent circuit. We have already proved that $I_n = 3I_0$. The zero sequence impedances of the high- and low-voltage windings are shown as Z_H and Z_L, respectively. The transformer impedance $Z_T = Z_H + Z_L$ on a per unit basis. This impedance is specified by the manufacturer as a percentage impedance on transformer MVA base, based on ONAN (natural liquid cooled rating, winding and core immersed on insulating liquid with fire point less than or equal to 300°C) for liquid immersed transformers or AA (natural air cooled, without forced ventilation for dry-type transformers) rating of the transformer. For example, a 138–13.8 kV transformer may be rated as follows:

40 MVA, ONAN ratings at 55°C rise

44.8 MVA, ONAN rating at 65°C rise

60 MVA, ONAF (forced air, i.e., fan cooled) rating at first stage of fan cooling, 65°C rise

75 MVA, ONAF second-stage fan cooling, 65°C rise

These ratings are normally applicable for an ambient temperature of 40°C, with an average of 30°C over a period of 24 h. The percentage impedance will be normally specified on a 40-MVA or possibly a 44.8-MVA base.

The percentage impedance we will change dependent on the tap settings on the transformer windings. The transformers may be provided with off-load or under-load tap changing [12].

The difference between the zero sequence impedance circuits of wye-wye connected shell- and core-form transformers in Table 2.1 is noteworthy. Connections 8 and 9 are for a core-type transformer and connections 7 and 10 are for a shell-type transformer. The impedance Z_M accounts for magnetic coupling between the phases of a core-type transformer.

2.5.5 Three-Winding Transformers

The theory of linear networks can be extended to apply to multi-winding transformers. A linear network having n terminals requires $\frac{1}{2}n(n + 1)$ quantities to specify it completely for a given frequency and EMF. Figure 2.7 shows the wye equivalent circuit of a three-winding transformer. One method to obtain the necessary data is to designate the pairs of terminals as $1, 2,\ldots, n$. All the terminals are then short-circuited except terminal one and a suitable EMF is applied across it. The current flowing in

TABLE 2.1 Equivalent Positive, Negative, and Zero Sequence Circuits of Two-Winding Transformers

No.	Winding Connections	Zero Sequence Circuit	Positive or Negative Sequence Circuit
1	Z_{nH}	H Z_H Z_L L; $3Z_{nH}$; N_0	H Z_H Z_L L; N_1 or N_2
2		H Z_H Z_L L; N_0	H Z_H Z_L L; N_1 or N_2
3		H Z_H Z_L L; N_0	H Z_H Z_L L; N_1 or N_2
4		H Z_H Z_L L; N_0	H Z_H Z_L L; N_1 or N_2
5		H Z_H Z_L L; N_0	H Z_H Z_L L; N_1 or N_2
6		H Z_H Z_L L; N_0	H Z_H Z_L L; N_1 or N_2
7		H Z_H Z_L L; N_0	H Z_H Z_L L; N_1 or N_2
8		H Z_H Z_L L; N_0 Z_M	H Z_H Z_L L; N_1 or N_2
9	Z_{nH} Z_{nL}	H Z_H $3Z_{nH}$ $3Z_{nL}$ Z_L L; Z_M; N_0	H Z_H Z_L L; N_1 or N_2
10	Z_{nH} Z_{nL}	H Z_H $3Z_{nH}$ $3Z_{nL}$ Z_L L; N_0	H Z_H Z_L L; N_1 or N_2

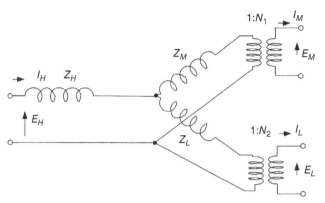

Figure 2.7   Wye equivalent circuit of a three-winding transformer.

each pair of terminals is measured. This is repeated for all the terminals. For a three-winding transformer:

$$Z_H = \frac{1}{2}(Z_{HM} + Z_{HL} - Z_{ML})$$

$$Z_M = \frac{1}{2}(Z_{ML} + Z_{HM} - Z_{HL}) \tag{2.13}$$

$$Z_L = \frac{1}{2}(Z_{HL} + Z_{ML} - Z_{HM})$$

where Z_{HM} = leakage impedance between the H and X windings, as measured on the H winding with M winding short-circuited and L winding open circuited; Z_{HL}= leakage impedance between the H and L windings, as measured on the H winding with L winding short-circuited and M winding open circuited; Z_{ML} = leakage impedance between the M and L windings, as measured on the M winding with L winding short-circuited and H winding open circuited.

Equation (2.13) can be written as

$$\begin{vmatrix} Z_H \\ Z_M \\ Z_L \end{vmatrix} = 1/2 \begin{vmatrix} 1 & 1 & -1 \\ 1 & -1 & 1 \\ -1 & 1 & 1 \end{vmatrix} \begin{vmatrix} Z_{HM} \\ Z_{HL} \\ Z_{ML} \end{vmatrix} \tag{2.14}$$

We also see that

$$\begin{aligned} Z_{HL} &= Z_H + Z_L \\ Z_{HM} &= Z_H + Z_M \\ Z_{ML} &= Z_M + Z_L \end{aligned} \tag{2.15}$$

Tables 2.1 and 2.2 clearly show that the positive and negative sequence impedances are the same irrespective of winding connections, but zero sequence impedance differs.

Example 2.2   A three-winding autotransformer is shown in Figure 2.8a. Construct the positive, negative, and zero sequence equivalent circuits and write their equations.

TABLE 2.2   Equivalent Positive, Negative, and Zero Sequence Circuits for Three-Winding Transformers

The equivalent T-circuit of positive or negative sequence impedances is shown in Figure 2.8b.

$$\begin{vmatrix} Z_H \\ Z_X \\ Z_Y \end{vmatrix} = \frac{1}{2} \begin{vmatrix} 1 & 1 & -1 \\ 1 & -1 & 1 \\ -1 & 1 & 1 \end{vmatrix} \begin{vmatrix} Z_{HX} \\ Z_{HY} \\ Z_{XY} \end{vmatrix}$$

Here all impedances are in pu.

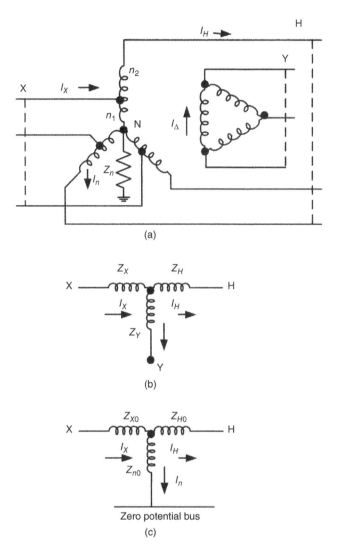

Figure 2.8   (a) Wye connected autotransformer equivalent circuit; (b, c) equivalent positive sequence and zero sequence circuits.

The zero sequence T-equivalent is shown in Figure 2.8c

$$
\begin{vmatrix} Z_{X0} \\ Z_{H0} \\ Z_{n0} \end{vmatrix} = \frac{1}{2} \begin{vmatrix} 1 & 1 & -1 & (n-1)/n \\ 1 & -1 & 1 & -(n-1)/n^2 \\ -1 & 1 & 1 & 1/n \end{vmatrix} \begin{vmatrix} Z_{HX} \\ Z_{HY} \\ Z_{XY} \\ 6Z_n \end{vmatrix}
$$

where n_1, n_2 are number of turns as shown in Figure 2.8a.

$$
n = \frac{n_1 + n_2}{n}
$$

2.6 EXAMPLE OF CONSTRUCTION OF SEQUENCE NETWORKS

Example 2.3   A single line diagram of a simple power system is shown in Figure 2.9. The positive, negative, and zero sequence data of the system components are shown in Table 2.3. It is required to construct positive, negative, and zero sequence networks and reduce these to single impedance.

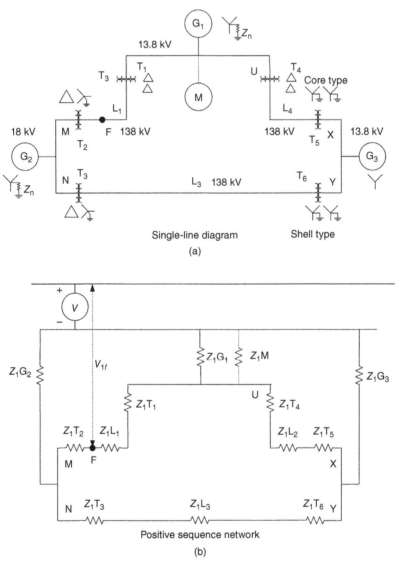

Single-line diagram

(a)

Positive sequence network

(b)

Figure 2.9   (a) A single-line diagram of a distribution system; (b–d) positive, negative, and zero sequence networks for a fault at point F.

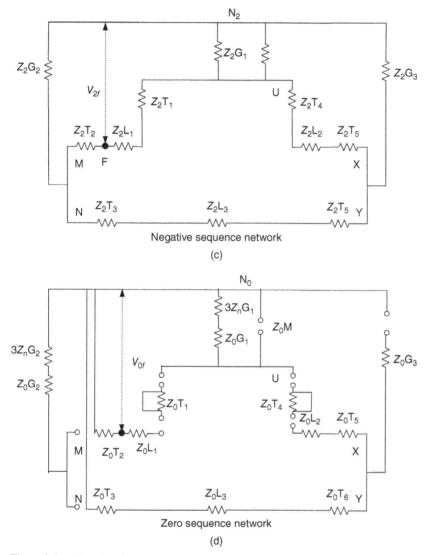

Figure 2.9   (*Continued*)

The positive sequence network is shown in Figure 2.9b. There are three generators in the system, and their positive sequence impedances are clearly marked in Figure 2.9b. The generator impedances are returned to a common bus. The Thévenin voltage at the fault point is shown to be equal to the generator voltages, which are all equal. This has to be so as all load currents are neglected, that is, all the shunt elements representing loads are open-circuited. Therefore, the voltage magnitudes and phase angles of all three generators must be equal. When load flow is considered, generation voltages will differ in magnitude and phase, and the voltage vector

TABLE 2.3   Rating and Impedance Data for Example

Equipment Designation	Equipment Description	Impedance Data (All in pu on equipment rating base)
G_1, G_2, and G_3	4-pole, 60 Hz, 13.8 kV synchronous generators, 100 MVA, 0.85 pf. G_1 and G_2 grounded through 400 A resistors, G_3 ungrounded	$X_d'' = 0.15$ $X_2 = 0.13$ $X_0 = 0.07$
Transformers T_1, T_2, T_3, T_4, and T_5	All transformers rated 50 MVA, each, winding connections as shown	$X_1 = X_2 = 0.09$ $XR = 31$ $Z_0 = 0.08$
L_1, L_2, and L_3	138 kV lines, each 10 miles long, 477 kcmil ACSR conductor, one 3#8 ground wire, conductor outside diameter = 0.858″, GMR = 0.029 ft, ground wire outside diameter = 0.277″, GMR = 0.00208 ft, GMD = 12.6′, earth resistivity = 100 Ω/m, tower footing resistance = 10 Ω	$Z_1 = Z_2 = 0.1961 + j0.7396$ Ω/mile, $Y = 5.8002$ μS/mile $Z_0 = 0.658 + j2.593$ Ω/mile, $Y = 3.097$ μS/mile
Motor M_1	10,000 hp, 4 pole, pf = 0.93, efficiency = 0.94	$Z_1 = 0.17$, $Z_2 = 0.15$, $Z_0 = \infty$ 100 MVA base

at the chosen fault point, prior to the fault, can be calculated based on load flow. However, the load currents are normally ignored in short-circuit calculations. Fault duties of switching devices are calculated based on rated system voltage rather than the actual voltage, which varies with load flow. This is generally true, unless the prefault voltage at the fault point remains continuously above or below the rated voltage.

Figure 2.9c shows the negative sequence network. Note the similarity with the positive sequence network with respect to interconnection of various system components.

Figure 2.9d shows zero sequence impedance network. This is based on the transformer zero sequence networks shown in Table 2.1 The neutral impedance is multiplied by a factor of three. Note that zero sequence impedance of motor is infinite. This is so because as per practice in the United States, the wye point of the motor windings is not grounded and left floating. In some European countries, the practice is to ground the neutrals of wye connected motor windings. A reader may carefully check the zero sequence impedance connections shown depending on the transformer-winding connections as illustrated in Table 2.1.

Let us now assign some numerical values to the positive, negative, and zero sequence impedances, as shown in Table 2.3. We will reduce each network to single impedance as seen from point F marked in Figure 2.9.

The equivalent circuit connections for the positive sequence network are shown in Figure 2.10 and its single impedance equivalent is 1.19 + j30.33 Ω. Wye-delta transformations of impedances are involved.

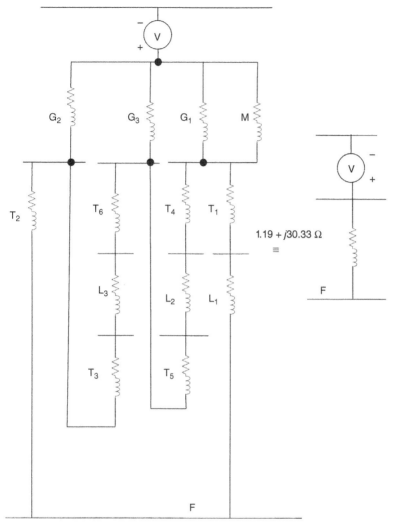

Figure 2.10   Equivalent circuit reduction for the positive sequence network for a fault at point F.

Referring to Figure 2.11, wye-to-delta and delta-to-wye impedance transformations are given by

$$Z_1 = \frac{Z_{12}Z_{31}}{Z_{12} + Z_{23} + Z_{31}}$$

$$Z_2 = \frac{Z_{12}Z_{23}}{Z_{12} + Z_{23} + Z_{31}} \qquad (2.16)$$

$$Z_3 = \frac{Z_{23}Z_{31}}{Z_{12} + Z_{23} + Z_{31}}$$

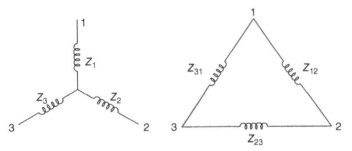

Figure 2.11   Wye-delta and delta-wye transformations of impedances.

and from wye to delta:

$$Z_{12} = \frac{Z_1 Z_2 + Z_2 Z_3 + Z_3 Z_1}{Z_3}$$

$$Z_{23} = \frac{Z_1 Z_2 + Z_2 Z_3 + Z_3 Z_1}{Z_1} \qquad (2.17)$$

$$Z_{31} = \frac{Z_1 Z_2 + Z_2 Z_3 + Z_3 Z_1}{Z_2}$$

Similarly, the negative sequence equivalent impedance is $1.15 + j\,29.01\ \Omega$. This differs slightly from the positive sequence impedance, mainly because of the lower negative sequence impedance of generators. Sometimes, for simplicity the negative sequence impedance is assumed equal to the positive sequence impedance even though the system under study may have considerable rotating machinery—this will introduce errors in the calculations.

The zero sequence impedance into point F is only that of transformer T_2, 8% on 50 MVA base. This gives $Z_0 = 30.52\ \Omega$. Considering an X/R ratio of 30.8 (Table 2.3), zero sequence impedance to the fault point is $0.99 + j30.49\ \Omega$.

Once we have reduced the impedances at the interested point, further calculations can proceed. For example, for a three-phase bolted fault we need to consider only positive sequence impedance and this gives a three-phase fault current of 2.626 kA at $-87.74°$. Also see References [8–12].

REFERENCES

[1] CF Wagner and RD Evans. *Symmetrical Components*. New York: McGraw-Hill, 1933.

[2] E Clarke. *Circuit Analysis of Alternating Current Power Systems*, vol. 1. New York: Wiley, 1943.

[3] JO Bird. *Electrical Circuit Theory and Technology*. Oxford, UK: Butterworth Heinemann, 1997.

[4] GW Stagg and A Abiad. *Computer Methods in Power Systems Analysis*. New York: McGraw-Hill, 1968.

[5] WE Lewis and DG Pryce. *The Application of Matrix Theory to Electrical Engineering*. London: E&FN Spon, 1965, chapter 6.

[6] LJ Myatt. *Symmetrical Components*. Oxford, UK: Pergamon Press, 1968.

[7] CA Worth (ed.). *J. & P. Transformer Book*, 11th edition. London: Butterworth, 1983.

[8] *Westinghouse Electric Transmission and Distribution Handbook*, 4th edition. East Pittsburgh, PA: Westinghouse Electric Corp., 1964.

[9] AE Fitzgerald, C Kingsley, and A Kusko. *Electric Machinery*, 3rd edition. New York: McGraw-Hill, 1971.

[10] CL Fortescu. Method of symmetrical coordinates applied to the solution of polyphase networks. *American Institute of Electrical Engineers*, vol. 37, pp. 1027–1140, 1918.

[11] IEEE Std. 399, IEEE Recommended Practice for Power Systems Analysis, 1997.

[12] JC Das. *Power System Analysis*, 2nd edition. Boca Raton, FL: CRC Press, 2012.

CHAPTER 3

SYMMETRICAL COMPONENTS-TRANSMISSION LINES AND CABLES

TRANSMISSION lines are modeled with respect to their lengths. For short lines less than approximately 80 km, the line capacitance is altogether ignored. For lengths 60–240 km, either Π or T model is used. For greater length, long lines, the distributed parameter representation of the line is necessary. Partial differential equations are used and the power transmission occurs like a wave motion.

We will not discuss

- The calculations and expressions of line resistance
- Inductance of a three-phase line
- Characteristic impedance
- Phase constant and attenuation constant
- Compensation of the lines with lumped elements like shunt capacitors or reactors
- Series compensated lines
- Reflection coefficients
- Lattice diagrams
- Surge impedance loading
- Infinite line
- Tuned power line
- Ferranti effect
- Symmetrical lines
- Circle diagrams
- Power flow over transmission lines
- Stability considerations
- Reactive power compensation
- ABCD constants of the lines

Understanding Symmetrical Components for Power System Modeling, First Edition. J.C. Das.
© 2017 by The Institute of Electrical and Electronics Engineers, Inc. Published 2017 by John Wiley & Sons, Inc.

TABLE 3.1 ABCD Constants of Transmission Lines

Line Length	Equivalent Circuit	A	B	C	D
Short	Series impedance only	1	Z	0	1
Medium	Nominal Π	$1 + \frac{1}{2}\,YZ$	Z	$Y[1 + 1/4(YZ)]$	$1 + \frac{1}{2}\,YZ$
Medium	Nominal T	$1 + \frac{1}{2}\,YZ$	$Z[1 + 1/4(YZ)]$	Y	$1 + \frac{1}{2}\,YZ$
Long	Distributed parameters	$\cosh \gamma\ 1$	$Z_0 \sinh \gamma\ 1$	$(\sinh \gamma\ 1)/Z_0$	$\cosh \gamma\ 1$

- Modal analysis
- Corona loss

References [1–18] may be seen. Table 3.1 shows the ABCD constants.

3.1 IMPEDANCE MATRIX OF THREE-PHASE SYMMETRICAL LINE

In Chapter 1, we decoupled a symmetrical three-phase line 3×3 matrix having equal self-impedances and mutual impedances (Equation 1.3).We showed that the off-diagonal elements of the sequence impedance matrix are zero. In high-voltage transmission lines which are transposed, this is generally true and the mutual couplings between phases are almost equal. However, the same cannot be said of distribution lines and these may have unequal off-diagonal terms. In many cases, the off-diagonal terms are smaller than the diagonal terms and the errors introduced in ignoring these will be small. Sometimes equivalence can be drawn by the equations:

$$Z_s = \frac{Z_{aa} + Z_{bb} + Z_{cc}}{3}$$

$$Z_m = \frac{Z_{ab} + Z_{bc} + Z_{ca}}{3} \tag{3.1}$$

that is, an average of the self- and mutual impedances can be taken. The sequence impedance matrix then has only diagonal terms, this is further illustrated with an example to follow.

3.2 THREE-PHASE LINE WITH GROUND CONDUCTORS

A three-phase transmission line has couplings between phase-to-phase conductors and also between phase-to-ground conductors. Consider a three-phase line with two ground conductors, as shown in Figure 3.1. The voltage V_a can be written as

$$\begin{aligned}
V_a &= R_a I_a + j\omega L_a I_a + j\omega L_{ab} I_b + j\omega L_{ac} I_c + j\omega L_{aw} I_w + j\omega L_{av} I_v \\
&\quad - j\omega L_{an} + V_a' + R_n I_n + j\omega L_n I_n - j\omega L_{an} I_a - j\omega L_{bn} I_b \\
&\quad - j\omega L_{cn} I_c - j\omega L_{wn} I_w - j\omega L_{vn} I_v
\end{aligned} \tag{3.2}$$

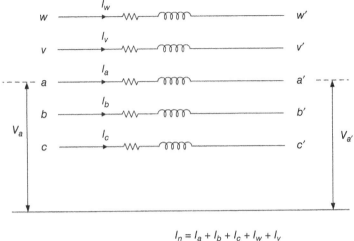

$$I_n = I_a + I_b + I_c + I_w + I_v$$

Figure 3.1   Transmission line section with two ground conductors.

where

$R_a, R_b, \ldots, R_n$ are resistances of phases $a, b, \ldots, n$
$L_a, L_b, \ldots, L_n$ are the self-inductances
$L_{ab}, L_{ac}, \ldots, L_{an}$ are the mutual inductances

This can be written as

$$
\begin{aligned}
V_a &= (R_a + R_n)I_a + R_n I_b + R_n I_c + j\omega(L_a + L_n - 2L_{an})I_a \\
&\quad + j\omega(L_{ab} + L_n - L_{an} - L_{bn})I_b + j\omega(L_{ac} + L_n - L_{an} - L_{cn})I_c + R_n I_w \\
&\quad + j\omega(L_{aw} + L_n - L_{an} - L_{wn})I_w + R_n I_v + j\omega(L_{av} + L_n - L_{an} - L_{vn})I_v + V'_a \\
&= Z_{aa-g}I_a + Z_{ab-g}I_b + Z_{ac-g}I_c + Z_{aw-g}I_w + Z_{av-g}I_v + V'_a
\end{aligned}
\tag{3.3}
$$

where $Z_{aa\text{-}g}$ and $Z_{hb\text{-}g}$ are the self-impedances of a conductor with ground return, and $Z_{ab\text{-}g}$ and $Z_{ac\text{-}g}$ are the mutual impedances between two conductors with common earth return. Similar equations apply to the voltages of other phases and ground wires. The following matrix then holds for the voltage differentials between terminals marked w, v, a, b, and c and w', v', a', b', and c':

$$
\begin{vmatrix}
\Delta V_a \\
\Delta V_b \\
\Delta V_c \\
\Delta V_w \\
\Delta V_v
\end{vmatrix}
=
\begin{vmatrix}
Z_{aa-g} & Z_{ab-g} & Z_{ac-g} & Z_{aw-g} & Z_{av-g} \\
Z_{ba-g} & Z_{bb-g} & Z_{bc-g} & Z_{bw-g} & Z_{bv-g} \\
Z_{ca-g} & Z_{cb-g} & Z_{cc-g} & Z_{cw-g} & Z_{cv-g} \\
Z_{wa-g} & Z_{wb-g} & Z_{wc-g} & Z_{ww-g} & Z_{wv-g} \\
Z_{va-g} & Z_{vb-g} & Z_{vc-g} & Z_{vw-g} & Z_{vv-g}
\end{vmatrix}
\begin{vmatrix}
I_a \\
I_b \\
I_c \\
I_w \\
I_v
\end{vmatrix}
\tag{3.4}
$$

In the partitioned form, this matrix can be written as

$$
\begin{vmatrix}
\Delta \bar{V}_{abc} \\
\Delta \bar{V}_{wv}
\end{vmatrix}
=
\begin{vmatrix}
\bar{Z}_A & \bar{Z}_B \\
\bar{Z}_C & \bar{Z}_D
\end{vmatrix}
\begin{vmatrix}
\bar{I}_{abc} \\
\bar{I}_{wv}
\end{vmatrix}
\tag{3.5}
$$

Considering that the ground wire voltages are zero,

$$\Delta \bar{V}_{abc} = \bar{Z}_A \bar{I}_{abc} + \bar{Z}_B \bar{I}_{wv}$$
$$0 = \bar{Z}_C \bar{I}_{abc} + \bar{Z}_D \bar{I}_{wv} \tag{3.6}$$

Thus,

$$\bar{I}_{wv} = -\bar{Z}_D^{-1} \bar{Z}_C \bar{I}_{abc} \tag{3.7}$$

$$\Delta \bar{V}_{abc} = (\bar{Z}_A - \bar{Z}_B \bar{Z}_D^{-1} \bar{Z}_C) \bar{I}_{abc} \tag{3.8}$$

This can be written as

$$\Delta \bar{V}_{abc} = \bar{Z}_{abc} \bar{I}_{abc} \tag{3.9}$$

$$\bar{Z}_{abc} = \bar{Z}_A - \bar{Z}_B \bar{Z}_D^{-1} \bar{Z}_C = \begin{vmatrix} Z_{aa'-g} & Z_{ab'-g} & Z_{ac'-g} \\ Z_{ba'-g} & Z_{bb'-g} & Z_{bc'-g} \\ Z_{ca'-g} & Z_{cb'-g} & Z_{cc'-g} \end{vmatrix} \tag{3.10}$$

The five-conductor circuit is reduced to an equivalent three-conductor circuit. The technique is applicable to circuits with any number of ground wires, provided that the voltages are zero in the lower portion of the voltage vector.

The final 3×3 matrix can be converted into positive, negative, and zero sequence matrix with the transformations discussed in Chapter 1.

Concept 3.1 *A transmission line matrix representing ground wires or bundle conductors can always be converted into a 3×3 matrix. It is practical to expect that the off-diagonal elements are not identical, especially for distribution lines, but the variations are small and these can be converted using (Equation 3.1).*

3.3 BUNDLE CONDUCTORS

Consider bundle conductors, consisting of two conductors per phase (Figure 3.2). The original circuit of conductors a, b, c and a', b', c' can be transformed into an equivalent conductor system of a'', b'', and c''.

Each conductor in the bundle carries a different current and has a different self- and mutual impedance because of its specific location. Let the currents in the conductors be I_a, I_b, and I_c and I'_a, I'_b, and I'_c, respectively. The following primitive matrix equation can be written as

$$\begin{vmatrix} V_a \\ V_b \\ V_c \\ V'_a \\ V'_b \\ V'_c \end{vmatrix} \begin{vmatrix} Z_{aa} & Z_{ab} & Z_{ac} & Z_{aa'} & Z_{ab'} & Z_{ac'} \\ Z_{ba} & Z_{bb} & Z_{bc} & Z_{ba'} & Z_{bb'} & Z_{bc'} \\ Z_{ca} & Z_{cb} & Z_{cc} & Z_{ca'} & Z_{cb'} & Z_{cc'} \\ Z_{a'a} & Z_{a'b} & Z_{a'c} & Z_{a'a'} & Z_{a'b'} & Z_{a'c'} \\ Z_{b'a} & Z_{b'b} & Z_{b'c} & Z_{b'a'} & Z_{b'b'} & Z_{b'c'} \\ Z_{c'a} & Z_{c'b} & Z_{c'c} & Z_{c'a'} & Z_{c'b'} & Z_{c'c'} \end{vmatrix} \begin{vmatrix} I_a \\ I_b \\ I_c \\ I_{a'} \\ I_{b'} \\ I_{c'} \end{vmatrix} \tag{3.11}$$

Transform to

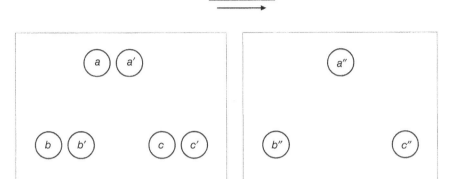

Figure 3.2   Transformation of bundle conductors to single conductors.

This can be partitioned so that

$$\left|\begin{array}{c} \bar{V}_{abc} \\ \bar{V}_{a'b'c'} \end{array}\right| = \left|\begin{array}{cc} \bar{Z}_1 & \bar{Z}_2 \\ \bar{Z}_3 & \bar{Z}_4 \end{array}\right| \left|\begin{array}{c} \bar{I}_{abc} \\ \bar{I}_{a'b'c'} \end{array}\right| \qquad (3.12)$$

for symmetrical arrangement of bundle conductors $\bar{Z}_1 = \bar{Z}_4$.

Modify so that the lower portion of the vector goes to zero. Assume that

$$\begin{aligned} V_a &= V'_a = V''_a \\ V_b &= V'_b = V''_b \\ V_c &= V'_c = V''_c \end{aligned} \qquad (3.13)$$

The upper part of the matrix can then be subtracted from the lower part:

$$\left|\begin{array}{c} V_a \\ V_b \\ V_c \\ 0 \\ 0 \\ 0 \end{array}\right| = \left|\begin{array}{cccccc} Z_{aa} & Z_{ab} & Z_{ac} & Z_{aa'} & Z_{ab'} & Z_{ac'} \\ Z_{ba} & Z_{bb} & Z_{bc} & Z_{ba'} & Z_{bb'} & Z_{bc'} \\ Z_{ca} & Z_{cb} & Z_{cc} & Z_{ca'} & Z_{cb'} & Z_{cc'} \\ Z_{a'a}-Z_{aa} & Z_{a'b}-Z_{ab} & Z_{a'c}-Z_{ac} & Z_{a'a'}-Z_{aa'} & Z_{a'b'}-Z_{ab'} & Z_{a'c'}-Z_{ac'} \\ Z_{b'a}-Z_{ba} & Z_{b'b}-Z_{bb} & Z_{b'c}-Z_{bc} & Z_{b'a'}-Z_{ba'} & Z_{b'b'}-Z_{bb'} & Z_{b'c'}-Z_{bc'} \\ Z_{c'a}-Z_{ca} & Z_{c'b}-Z_{cb} & Z_{c'c}-Z_{cc} & Z_{c'a'}-Z_{ca'} & Z_{c'b'}-Z_{cb'} & Z_{c'c'}-Z_{cc'} \end{array}\right| \left|\begin{array}{c} I_a \\ I_b \\ I_c \\ I_{a'} \\ I_{b'} \\ I_{c'} \end{array}\right|$$
$$(3.14)$$

We can write it in the partitioned form as

$$\left|\begin{array}{c} \bar{V}_{abc} \\ 0 \end{array}\right| = \left|\begin{array}{cc} \bar{Z}_1 & \bar{Z}_2 \\ \bar{Z}_2^t - \bar{Z}_1 & \bar{Z}_4 - \bar{Z}_2 \end{array}\right| \left|\begin{array}{c} \bar{I}_{abc} \\ \bar{I}_{a'b'c'} \end{array}\right| \qquad (3.15)$$

$$\begin{aligned} I''_a &= I_a + I'_a \\ I''_b &= I_b + I'_b \\ I''_c &= I_c + I'_c \end{aligned} \qquad (3.16)$$

The matrix is modified as shown below:

$$
\begin{vmatrix}
Z_{aa} & Z_{ab} & Z_{ac} & Z_{aa'} - Z_{aa} & Z_{ab'} - Z_{ab} & Z_{ac'} - Z_{ac} \\
Z_{ba} & Z_{bb} & Z_{bc} & Z_{ba'} + Z_{ba} & Z_{bb'} + Z_{bb} & Z_{bc'} - Z_{bc} \\
Z_{ca} & Z_{cb} & Z_{cc} & Z_{ca'} - Z_{ca} & Z_{cb'} - Z_{cb} & Z_{cc'} - Z_{cc} \\
Z_{a'a} - Z_{aa} & Z_{a'b} - Z_{ab} & Z_{a'c} - Z_{ac} & Z_{a'a'} - Z_{aa'} - Z_{a'a} + Z_{aa} & Z_{a'b'} - Z_{ab'} - Z_{a'b} + Z_{ab} & Z_{a'c'} - Z_{ac'} - Z_{a'c} + Z_{ac} \\
Z_{b'a} - Z_{ba} & Z_{b'b} - Z_{bb} & Z_{b'c} - Z_{bc} & Z_{b'a'} - Z_{ba'} - Z_{b'a} + Z_{ba} & Z_{b'b'} - Z_{bb'} - Z_{b'b} + Z_{bb} & Z_{b'c'} - Z_{bc'} - Z_{b'c} + Z_{bc} \\
Z_{c'a} - Z_{ca} & Z_{c'b} - Z_{cb} & Z_{c'c} - Z_{cc} & Z_{c'a'} - Z_{ca'} - Z_{c'a} + Z_{ca} & Z_{c'b'} - Z_{cb'} - Z_{c'b} + Z_{cb} & Z_{c'c'} - Z_{cc'} - Z_{c'c} + Z_{cc}
\end{vmatrix}
\begin{vmatrix}
I_a + I'_a \\
I_b + I'_b \\
I_c + I'_c \\
I'_a \\
I'_b \\
I'_c
\end{vmatrix}
$$

$$(3.17)$$

or in partitioned form:

$$
\begin{vmatrix} \bar{V}_{abc} \\ 0 \end{vmatrix} = \begin{vmatrix} \bar{Z}_1 & \bar{Z}_2 - \bar{Z}_1 \\ \bar{Z}_2^t - \bar{Z}_1 & (\bar{Z}_4 - \bar{Z}_2) - (\bar{Z}_2^t - \bar{Z}_1) \end{vmatrix} \begin{vmatrix} \bar{I}_{abc} \\ \bar{I}_{a'b'c''} \end{vmatrix} \qquad (3.18)
$$

This can now be reduced to the following 3×3 matrix as before:

$$
\begin{vmatrix} V''_a \\ V''_b \\ V''_c \end{vmatrix} = \begin{vmatrix} Z''_{aa} & Z''_{ab} & Z''_{ac} \\ Z''_{ba} & Z''_{bb} & Z''_{bc} \\ Z''_{ca} & Z''_{cb} & Z''_{cc} \end{vmatrix} \begin{vmatrix} I''_a \\ I''_b \\ I''_c \end{vmatrix} \qquad (3.19)
$$

3.4   CARSON'S FORMULA

The theoretical value of $Z_{abc\text{-}g}$ can be calculated by Carson's formula (c. 1926). This is of importance even today in calculations of line constants. For an n-conductor configuration, the earth is assumed as an infinite uniform solid with a constant resistivity. Figure 3.3 shows image conductors in the ground at a distance equal to the height of the conductors above ground and exactly in the same formation, with the same spacing between the conductors. A flat conductor formation is shown in Figure 3.3.

$$
Z_{ii} = R_i + 4\omega P_{ii} G + j \left[X_i + 2\omega G \ln \frac{S_{ii}}{r_i} + 4\omega Q_{ii} G \right] \Omega/\text{mile} \qquad (3.20)
$$

$$
Z_{ij} = 4\omega P_{ii} G + j \left[2\omega G \ln \frac{S_{ij}}{D_{ij}} + 4\omega Q_{ij} G \right] \Omega/\text{mile} \qquad (3.21)
$$

where

Z_{ii} = the self-impedance of conductor i with earth return (Ω/mile)

Z_{ij} = mutual impedance between conductors i and j (Ω/mile)

R_i = resistance of conductor in Ω/mile

S_{ii} = conductor to image distance of the ith conductor to its own image

S_{ij} = conductor to image distance of the ith conductor to the image of the jth conductor

D_{ij} = distance between conductors i and j

r_i = radius of conductor (ft)

ω = angular frequency

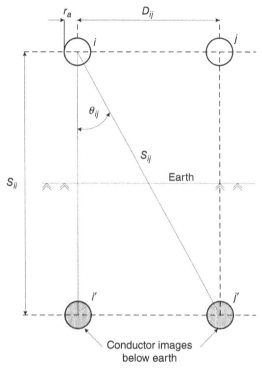

Figure 3.3   Conductors and their images in the earth, Carson's formula.

$G = 0.1609347 \times 10^{-7}$ Ω-cm

GMR_i = geometric mean radius of conductor i

ρ = soil resistivity

θ_{ij} = angle as shown in Figure 3.3

Expressions for P and Q are

$$P = \frac{\pi}{8} - \frac{1}{3\sqrt{2}}k\cos\theta + \frac{k^2}{16}\cos 2\theta \left(0.6728 + \ln\frac{2}{k}\right) + \frac{k^2}{16}\theta\sin\theta + \frac{k^3\cos 3\theta}{45\sqrt{2}}$$
$$- \frac{\pi k^4\cos 4\theta}{1536} \tag{3.22}$$

$$Q = -0.0386 + \frac{1}{2}\ln\frac{2}{k} + \frac{1}{3\sqrt{2}}\cos\theta - \frac{k^2\cos 2\theta}{64} + \frac{k^3\cos 3\theta}{45\sqrt{2}}$$
$$- \frac{k^4\sin 4\theta}{384} - \frac{k^4\cos 4\theta}{384}\left(\ln\frac{2}{k} = 1.0895\right) \tag{3.23}$$

where

$$k = 8.565 \times 10^4 S_{ij}\sqrt{f/\rho} \tag{3.24}$$

S_{ij} is in feet and ρ is soil resistivity in Ω-m, and f is the system frequency. This shows dependence on frequency as well as on soil resistivity.

3.4.1 Approximations to Carson's Equations

These approximations involve P and Q and the expressions are given by

$$P_{ij} = \frac{\pi}{8} \tag{3.25}$$

$$Q_{ij} = -0.03860 + \frac{1}{2} \ln \frac{2}{k_{ij}} \tag{3.26}$$

Using these assumptions, $f = 60$ Hz and soil resistivity $= 100$ Ω-m, the equations reduce to

$$Z_{ii} = R_i + 0.0953 + j0.12134 \left(\ln \frac{1}{\text{GMR}_i} + 7.93402 \right) \Omega/\text{mile} \tag{3.27}$$

$$Z_{ij} = 0.0953 + j0.12134 \left(\ln \frac{1}{D_{ij}} + 7.93402 \right) \Omega/\text{mile} \tag{3.28}$$

Equations (3.27) and (3.28) are of practical significance for calculations of line impedances.

(Equations are not available in SI units.)

Concept 3.2 *Carson's Formula is important and practically all texts on transmission lines describe this in one form or the other. It facilitates calculations of transmission line parameters, and is practically indispensable for this effort.*

Example 3.1 Consider an unsymmetrical overhead line configuration, as shown in Figure 3.4. The phase conductors consist of 556.5 kcmil (556,500 circular mils) of ACSR conductor consisting of 26 strands of aluminum, two layers, and seven strands of steel. From the properties of ACSR conductor tables, the conductor has a resistance of 0.1807 Ω at 60 Hz and its GMR is 0.0313 ft at 60 Hz; conductor diameter = 0.927 in. The neutral consists of 336.4 KCMIL, ACSR conductor, resistance 0.259 Ω/mile at 60 Hz and 50°C and GMR 0.0278 ft, and conductor diameter 0.806 in. It is required to form a primitive Z matrix, convert it into a 3×3 Z_{abc} matrix, and then to sequence impedance matrix Z_{012}.

Using Equations (3.27) and (3.28)

$$Z_{aa} = Z_{bb} = Z_{cc} = 0.2760 + j1.3831$$
$$Z_{nn} = 0.3543 + j1.3974$$
$$Z_{ab} = Z_{ba} = 0.0953 + j0.8515$$
$$Z_{bc} = Z_{cb} = 0.053 + j0.7654$$
$$Z_{ca} = Z_{ac} = 0.0953 + j0.7182$$
$$Z_{an} = Z_{na} = 0.0953 + j0.7539$$
$$Z_{bn} = Z_{nb} = 0.0953 + j0.7674$$
$$Z_{cn} = Z_{nc} = 0.0953 + j0.7237$$

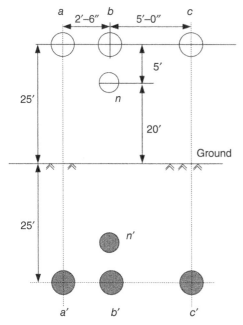

Figure 3.4 Distribution line configuration for calculations of line parameters.

Therefore, the primitive impedance matrix is

$$
\bar{Z}_{\mathrm{prim}} = \begin{vmatrix}
0.2760 + j1.3831 & 0.0953 + j0.8515 & 0.0953 + j0.7182 & 0.0953 + j0.7539 \\
0.0953 + j0.8515 & 0.2760 + j1.3831 & 0.0953 + j0.7624 & 0.0953 + j0.7674 \\
0.0953 + j0.7182 & 0.0953 + j0.7624 & 0.2760 + j1.3831 & 0.0953 + j0.7237 \\
0.0953 + j0.7539 & 0.0953 + j0.7674 & 0.0953 + j0.7237 & 0.3543 + j1.3974
\end{vmatrix}
$$

Eliminate the last row and column using Equation (3.10):

$$
\bar{Z}_{abc} = \begin{vmatrix}
0.2747 + j0.9825 & 0.0949 + j0.4439 & 0.0921 + j0.3334 \\
0.0949 + j0.4439 & 0.2765 + j0.9683 & 0.0929 + j0.3709 \\
0.0921 + j0.3334 & 0.0929 + j0.3709 & 0.2710 + j1.0135
\end{vmatrix}
$$

A reader can verify this transformation. Convert to sequence components impedances Z_{012} by using the transformation equation (see Chapter 1).

$$
\bar{Z}_{012} = \begin{vmatrix}
0.4606 + j1.7536 & 0.0194 + j0.0007 & -0.0183 + j0.0055 \\
-0.0183 + j0.0055 & 0.1808 + j0.6054 & -0.0769 - j0.0146 \\
0.0194 + j0.0007 & 0.0767 - j0.0147 & 0.1808 + j0.6054
\end{vmatrix}
$$

This shows the mutual coupling between sequence impedances. We could average out the self- and mutual impedances according to Equation (3.1):

$$Z_s = \frac{Z_{aa} + Z_{bb} + Z_{cc}}{3} = 0.2741 + j0.9973$$

$$Z_m = \frac{Z_{ab} + Z_{bc} + Z_{ca}}{3} = -0.0127 - j0.0044$$

The matrix Z_{abc} then becomes

$$\bar{Z}_{abc} = \begin{vmatrix} 0.2741 + j0.9973 & -0.0127 - j0.0044 & -0.0127 - j0.0044 \\ -0.0127 - j0.0044 & 0.2741 + j0.9973 & -0.0127 - j0.0044 \\ -0.0127 - j0.0044 & -0.0127 - j0.0044 & 0.2741 + j0.9973 \end{vmatrix} \quad \Omega/\text{mile}$$

and this gives

$$\bar{Z}_{012} = \begin{vmatrix} 0.2486 + j0.9885 & 0 & 0 \\ 0 & 0.2867 + j1.0017 & 0 \\ 0 & 0 & 0.2867 + j1.0017 \end{vmatrix} \quad \Omega/\text{mile}$$

Example 3.2   Figure 3.5 shows a high-voltage line with two 636,000 mils ACSR bundle conductors per phase. Conductor GMR = 0.0329 ft, resistance = 0.1688 Ω/mile, diameter = 0.977 in., and spacing are as shown in Figure 3.5. Calculate the primitive impedance matrix and reduce it to a 3 × 3 matrix, then convert it into a sequence component matrix.

The primitive matrix is 6 × 6 given by Equation (3.11) formed by partitioned matrices according to Equation (3.12). Thus, from $\bar{Z}_1$ and $\bar{Z}_2$ the primitive matrix can be written.

From Equations (3.27) and (3.28) and the specified spacing in Figure 3.5, matrix Z_1 is

$$\bar{Z}_1 = \begin{vmatrix} 0.164 + j1.3770 & 0.0953 + j0.5500 & 0.0953 + j0.4659 \\ 0.0953 + j0.5500 & 0.164 + j1.3770 & 0.0953 + j0.5500 \\ 0.0953 + j0.4659 & 0.0953 + j0.5500 & 0.164 + j1.3770 \end{vmatrix}$$

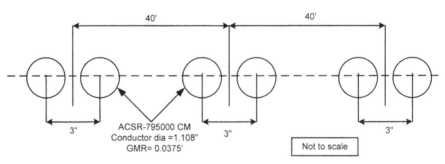

Figure 3.5   Configuration of bundle conductors, Example 3.2.

This is also equal to Z_4, as the bundle conductors are identical and symmetrically spaced. Matrix Z_2 of Equation (3.12) is

$$\bar{Z}_2 = \begin{vmatrix} 0.0953 + j0.8786 & 0.0953 + j0.5348 & 0.0953 + j0.4581 \\ 0.0953 + j0.5674 & 0.0953 + j0.8786 & 0.0953 + j0.5348 \\ 0.0953 + j0.4743 & 0.0953 + j0.8786 & 0.0953 + j0.8786 \end{vmatrix}$$

From these two matrices, we will calculate

$$\bar{Z}_1 - \bar{Z}_2 = \begin{vmatrix} 0.069 + j0.498 & j0.0150 & j0.0079 \\ -j0.0171 & 0.069 + j0.498 & j0.0150 \\ -j0.00847 & -j0.0170 & 0.069 + j0.498 \end{vmatrix}$$

and

$$\bar{Z}_k = (\bar{Z}_1 - \bar{Z}_2) - (\bar{Z}_2' - \bar{Z}_1) = \begin{vmatrix} 0.138 + j0.997 & -j0.0022 & -j0.0005 \\ -j0.0022 & 0.138 + j0.997 & -j0.0022 \\ -j0.0005 & -j0.0022 & 0.138 + j0.997 \end{vmatrix}$$

The inverse is

$$\bar{Z}_k^{-1} = \begin{vmatrix} 0.136 - j0.984 & 0.000589 - j0.002092 & 0.0001357 - j0.0004797 \\ 0.0005891 - j0.002092 & 0.136 - j0.981 & 0.0005891 - j0.002092 \\ 0.0001357 - j0.0004797 & 0.0005891 - j0.002092 & 0.136 - j0.981 \end{vmatrix}$$

then, the matrix $(\bar{Z}_2 - \bar{Z}_1)\bar{Z}_k^{-1}(\bar{Z}_2' - \bar{Z}_1)$ is

$$\begin{vmatrix} 0.034 + j0.2500 & -0.000018 - j0.000419 & 0.0000363 - j0.0003871 \\ -0.000018 - j0.000419 & 0.034 + j0.2500 & -0.000018 - j0.000419 \\ 0.0000363 - j0.000387 & -0.000018 - j0.000419 & 0.034 + j0.2500 \end{vmatrix}$$

Note that the off-diagonal elements are relatively small as compared to the diagonal elements. The required 3×3 transformed matrix is then Z_1 minus the above matrix:

$$\bar{Z}_{transformed} = \begin{vmatrix} 0.13 + j1.127 & 0.095 + j0.55 & 0.095 + j0.466 \\ 0.095 + j0.55 & 0.13 + j1.127 & 0.095 + j0.55 \\ 0.095 + j0.466 & 0.095 + j0.55 & 0.13 + j1.127 \end{vmatrix} \Omega/\text{mile}$$

Then the sequence impedance matrix is

$$\bar{Z}_{012} = \begin{vmatrix} 0.32 + j2.171 & 0.024 - j0.014 & -0.024 - j0.014 \\ -0.024 - j0.014 & 0.035 + j0.605 & -0.048 + j0.028 \\ 0.024 - j0.014 & 0.048 + j0.028 & 0.035 + j0.605 \end{vmatrix} \Omega/\text{mile}$$

3.5 CAPACITANCE OF LINES

The shunt capacitance per unit length of a two-wire, single-phase transmission line is

$$C = \frac{\pi \varepsilon_0}{\ln(D/r)} \text{F/m (Farads per meter)} \qquad (3.29)$$

where ε_0 is the permittivity of free space $= 8.854 \times 10^{-12}$ F/m, and other symbols are as defined before. For a three-phase line with equilaterally spaced conductors, the line-to-neutral capacitance is

$$C = \frac{2\pi \varepsilon_0}{\ln(D/r)} \text{F/m} \qquad (3.30)$$

For unequal spacing, D is replaced with GMD (Geometric Mean Distance). The capacitance is affected by the ground and the effect is simulated by a mirror image of the conductors exactly at the same depth as the height above the ground. These mirror-image conductors carry charges which are of opposite polarity to conductors above the ground (Figure 3.6). From this figure, the capacitance to ground is

$$C_n = \frac{2\pi \varepsilon_0}{\ln(GMD/r) - \ln\left(\sqrt[3]{S_{ab'}S_{bc'}S_{ca'}}\big/\sqrt[3]{S_{aa'}S_{bb'}S_{cc'}}\right)} \qquad (3.31)$$

This can be written as

$$C_n = \frac{2\pi \varepsilon_0}{\ln(D_m/D_s)} = \frac{10^{-9}}{18\ln(D_m/D_s)} F/m \qquad (3.32)$$

3.5.1 Capacitance Matrix

The capacitance matrix of a three-phase line is

$$\bar{C}_{abc} = \begin{vmatrix} C_{aa} & -C_{ab} & -C_{ac} \\ -C_{ba} & C_{bb} & -C_{bc} \\ -C_{ca} & -C_{cb} & C_{cc} \end{vmatrix} \qquad (3.33)$$

This is diagrammatically shown in Figure 3.7a. The capacitance between the phase conductor a and b is C_{ab}, and the capacitance between conductor a and ground is $C_{aa} - C_{ab} - C_{ac}$. If the line is perfectly symmetrical, all the diagonal elements are the same and all off-diagonal elements of the capacitance matrix are identical:

$$\bar{C}_{abc} = \begin{vmatrix} C & -C' & -C' \\ -C' & C & -C' \\ -C' & -C' & C \end{vmatrix} \qquad (3.34)$$

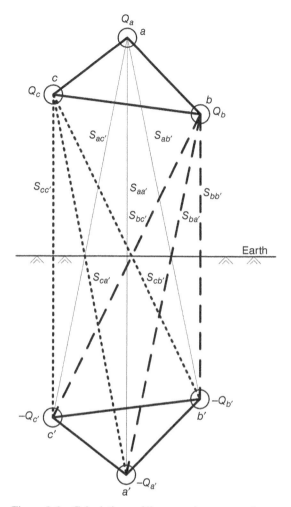

Figure 3.6   Calculations of line capacitances, conductors mirror images, spacing, and charges.

Symmetrical component transformation is used to diagonalize the matrix

$$\bar{C}_{012} = \bar{T}_s^{-1} \bar{C}_{abc} \bar{T}_s = \begin{vmatrix} C - 2C' & 0 & 0 \\ 0 & C + C' & 0 \\ 0 & 0 & C + C' \end{vmatrix}$$ (3.35)

The zero, positive, and negative sequence networks of capacitance of a symmetrical transmission line are shown in Figure 3.7b. The eigenvalues are $C - 2C'$, $C + C'$, and $C + C'$. The capacitance $C + C'$ can be written as $3C' + (C - 2C')$, that is, it is equivalent to the line capacitance of a three-conductor system plus the line-to-ground capacitance of a three-conductor system.

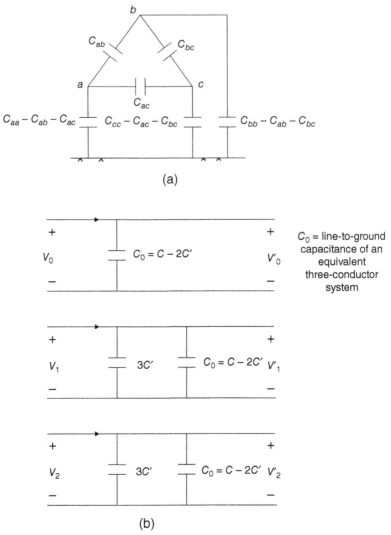

Figure 3.7   (a) Capacitances of a three-phase line; (b) equivalent positive, negative, and zero sequence networks of capacitances.

In a capacitor, $V = Q/C$. The capacitance matrix can be written as

$$\bar{V}_{abc} = \bar{P}_{abc}\bar{Q}_{abc} = \bar{C}_{abc}^{-1}\bar{Q}_{abc} \tag{3.36}$$

where $\bar{P}$ is called the *potential coefficient matrix*, that is,

$$\begin{vmatrix} V_a \\ V_b \\ V_c \end{vmatrix} = \begin{vmatrix} P_{aa} & P_{ab} & P_{ac} \\ P_{ba} & P_{bb} & P_{bc} \\ P_{ca} & P_{cb} & P_{cc} \end{vmatrix} \begin{vmatrix} Q_a \\ Q_b \\ Q_c \end{vmatrix} \tag{3.37}$$

where

$$P_{ii} = \frac{1}{2\pi\varepsilon_0} \ln \frac{S_{ii}}{r_i} = 11.17689 \ln \frac{S_{ii}}{r_i} \tag{3.38}$$

$$P_{ij} = \frac{1}{2\pi\varepsilon_0} \ln \frac{S_{ij}}{D_{ij}} = 11.17689 \ln \frac{S_{ij}}{D_{ij}} \tag{3.39}$$

where

S_{ij} = conductor-to-image distance below ground (ft)

D_{ij} = conductor-to-conductor distance (ft)

r_i = radius of the conductor (ft)

ε_0 = permittivity of the medium surrounding the conductor = 1.424×10^{-8}

For sine-wave voltage and charge, the equation can be expressed as

$$\begin{vmatrix} I_a \\ I_b \\ I_c \end{vmatrix} = j\omega \begin{vmatrix} C_{aa} & -C_{ab} & -C_{ac} \\ -C_{ba} & -C_{bb} & -C_{bc} \\ -C_{ca} & -C_{cb} & C_{cc} \end{vmatrix} \begin{vmatrix} V_a \\ V_b \\ V_c \end{vmatrix} \tag{3.40}$$

The capacitance of three-phase lines with ground wires and with bundle conductors can be addressed as in the calculations of inductances. The primitive P matrix can be partitioned and reduces to a 3×3 matrix.

Example 3.3 Calculate the matrices P and C for Example 3.1. The neutral is 20 ft (6.09 m) above ground and the configuration of Figure 3.4 is applicable.

The mirror images of the conductors are drawn in Figure 3.4. This facilitates calculation of the spacing required in Equations (3.38) and (3.39) for the P matrix. Based on the geometric distances and conductor diameter the primitive P matrix is

$$\bar{P} = \begin{vmatrix} P_{aa} & P_{ab} & P_{ac} & P_{an} \\ P_{ba} & P_{bb} & P_{bc} & P_{bn} \\ P_{ca} & P_{cb} & P_{cc} & P_{cn} \\ P_{na} & P_{nb} & P_{nc} & P_{nn} \end{vmatrix}$$

$$= \begin{vmatrix} 80.0922 & 33.5387 & 21.4230 & 23.3288 \\ 33.5387 & 80.0922 & 25.7913 & 24.5581 \\ 21.4230 & 25.7913 & 80.0922 & 20.7547 \\ 23.3288 & 24.5581 & 20.7547 & 79.1615 \end{vmatrix}$$

This is reduced to a 3×3 matrix

$$P = \begin{vmatrix} 73.2172 & 26.3015 & 15.3066 \\ 26.3015 & 72.4736 & 19.3526 \\ 15.3066 & 19.3526 & 74.6507 \end{vmatrix}$$

Therefore, the required $\bar{C}$ matrix is inverse of $\bar{P}$, and $\bar{Y}_{abc}$ is

$$\bar{Y}_{abc} = j\omega\bar{P}^{-1} = \begin{vmatrix} j6.0141 & -j1.9911 & -j0.7170 \\ -j1.9911 & j6.2479 & -j1.2114 \\ -j0.7170 & -j1.2114 & j5.5111 \end{vmatrix} \mu \text{ siemens/mile}$$

3.6 CABLE CONSTANTS

The construction of cables varies widely; it is mainly a function of insulation type, method of laying, and voltage of application. For high-voltage applications above 230 kV, oil-filled paper insulated cables are used, though recent trends see the development of solid dielectric cables up to 345 kV. A three-phase solid dielectric cable has three conductors enclosed within a sheath and because the conductors are much closer to each other than those in an overhead line and the permittivity of insulating medium is much higher than that of air, the shunt capacitive reactance is much higher (upto 10 times or more) as compared to an overhead line. Thus, use of a T or Π model is required even for shorter cable lengths [19–28].

The inductance per unit length of a single conductor cable is given by

$$L = \frac{\mu_0}{2\pi} \ln \frac{r_1}{r_2} H/m \tag{3.41}$$

where r_1 is the radius of the conductor and r_2 is the radius of the sheath, that is, the cable outside diameter divided by 2.

When single-conductor cables are installed in magnetic conduits the reactance may increase by a factor of 1.5. Reactance is also dependent on conductor shape, that is, circular or sector, and on the magnetic binders in three-conductor cables.

3.6.1 Zero Sequence Impedance of the OH lines and Cables

The zero sequence impedance of the lines and cables is dependent upon the current flow through a conductor and return through the ground or sheaths and encounters the impedance of these paths. The zero sequence current flowing in one phase also encounters the currents arising out of that conductor self-inductance, from mutual inductance to other two phase conductors, from the mutual inductance to the ground and sheath return paths, and the self-inductance of the return paths. As an example, the zero sequence impedance of a three-conductor cable with a solidly bonded and grounded sheath is given by

$$z_0 = r_c + r_e + j0.8382 \frac{f}{60} \log_{10} \frac{D_e}{GMR_{3c}} \tag{3.42}$$

where

r_c = ac resistance of one conductor Ω/mile

r_e = ac resistance of earth return (depending upon equivalent depth of earth return, soil resistivity, taken as 0.286 Ω/mile)

D_e = distance to equivalent earth path [15]

GMR_{3c} = geometric mean radius of conducting path made up of three actual conductors taken as a group(in inches):

$$GMR_{3c} = \sqrt[3]{GMR_{1c} S^2} \tag{3.43}$$

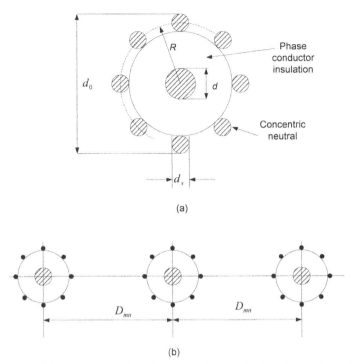

Figure 3.8   (a) construction of a concentric neutral cable; (b) configuration for calculations of series inductance.

where GMR_{1c} is the geometric mean radius of individual conductor and $S = (d+2t)$, where d = diameter of the conductor and t is the thickness of the insulation.

3.6.2 Concentric Neutral Underground Cable

We will consider a concentric neutral construction as shown in Figure 3.8a. The neutral is concentric to the conductor and consists of a number of copper strands that are wound helically over the insulation. Such cables are used for underground distribution, directly buried or installed in ducts. Referring to Figure 3.8a, d is the diameter of the conductor, d_0 is the outside diameter of the cable over the concentric neutral strands, and d_s is the diameter of an individual neutral strand. Three cables in flat formation are shown in Figure 3.8b. The GMR of a phase conductor and a neutral strand are given by the expression:

$$\text{GMR}_{cn} = \sqrt[n]{\text{GMR}_s n R^{n-1}} \qquad (3.44)$$

where GMR_{cn} is the equivalent GMR of the concentric neutral, GMR_s is the GMR of a single neutral strand, n is the number of concentric neutral strands, and R is the radius of a circle passing through the concentric neutral strands (Figure 3.8a) = $(d_0 - d_s)/2$ (ft).

The resistance of the concentric neutral is equal to the resistance of a single strand divided by the number of strands.

The geometric mean distance between concentric neutral and adjacent phase conductors is

$$D_{ij} = \sqrt[n]{D_{mn}^n - R^n} \tag{3.45}$$

where D_{ij} is the *equivalent* center-to-center distance of the cable spacing. Note that it is less than D_{mn}, the center-to-center spacing of the adjacent conductors (Figure 3.8b). Carson's formula can be applied and the calculations are similar to those in Example 3.1

Example 3.4   A concentric neutral cable system for 13.8 kV has a center-to-center spacing of 8 in. The cables are 500 kcmil, with 16 strands of #12 copper wires. The following data are supplied by the manufacturer:

GMR phase conductor = 0.00195 ft
GMR of neutral strand = 0.0030 ft
Resistance of phase conductor = 0.20 Ω/mile
Resistance of neutral strand = 10.76 Ω/mile. Therefore, the resistance of the concentric neutral = 10.76/16 = 0.6725 Ω/mile.
Diameter of neutral strand = 0.092 in.
Overall diameter of cable = 1.490 in.
Therefore, $R = (1.490 - 0.092)/24 = 0.0708$ ft
The effective conductor phase-to-phase spacing is approximately 8 in., from Equation (3.45).
The primitive matrix is a 6×6 matrix. In the partitioned form, Equation (3.12), the matrices are

$$\bar{Z}_A = \begin{vmatrix} 0.2953 + j1.7199 & 0.0953 + j1.0119 & 0.0953 + j0.9278 \\ 0.0953 + j1.0119 & 0.2953 + j1.7199 & 0.0953 + j1.0119 \\ 0.0953 + j0.9278 & 0.0953 + j1.0119 & 0.2953 + j1.7199 \end{vmatrix}$$

The spacing between the concentric neutral and the phase conductors is approximately equal to the phase-to-phase spacing of the conductors. Therefore,

$$\bar{Z}_B = \begin{vmatrix} 0.0953 + j1.284 & 0.0953 + j1.0119 & 0.0953 + j0.9278 \\ 0.0953 + j1.0119 & 0.0953 + j1.284 & 0.0953 + j1.0119 \\ 0.0953 + j0.9278 & 0.0953 + j1.0119 & 0.0953 + j1.284 \end{vmatrix}$$

Matrix $\bar{Z}_c = \bar{Z}_B$ and matrix $\bar{Z}_D$ is given by

$$\bar{Z}_D = \begin{vmatrix} 0.7678 + j1.2870 & 0.0953 + j1.0119 & 0.0953 + j0.9278 \\ 0.0953 + j1.0119 & 0.7678 + j1.2870 & 0.0953 + j1.0119 \\ 0.0953 + j0.9278 & 0.0953 + j1.0119 & 0.7678 + j1.2870 \end{vmatrix} z$$

This primitive matrix can be reduced to a 3×3 matrix, as in other examples and then to Z_{012}.

3.6.3 Capacitance of Cables

In a single-conductor cable, the capacitance per unit length is given by

$$C = \frac{2\pi\varepsilon\varepsilon_0}{\ln(r_1/r_2)} \text{F/m} \qquad (3.46)$$

By change of units, this can be written as

$$C = \frac{7.35\varepsilon}{\log(r_1/r_2)} \text{pF/ft} \qquad (3.47)$$

Note that ε is the permittivity of the dielectric medium relative to air. The capacitances in a three-conductor cable are shown in Figure 3.9. This assumes a symmetrical construction and the capacitances between conductors and from conductors to the sheath is equal. The circuit of Figure 3.9a is successively transformed and Figure 3.9d shows that the net capacitance per phase $= C_1 + 3C_2$.

This gives the capacitance of a single-conductor shielded cable. Table 3.2 gives values of ε for various cable insulation types.

Concept 3.3   *The models discussed above can be used for power frequencies. For the power system components it is often necessary to construct models good for higher frequencies up to VHF (very high frequencies of the order of 100 kHz to 50 MHz (e.g., for GIS). A single model cannot be sufficient and the frequency-dependent models are available depending on the study to be undertaken.*

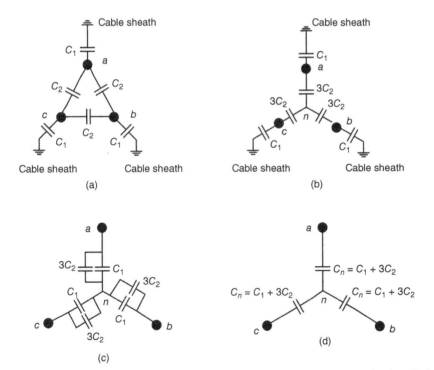

Figure 3.9   (a) Capacitance of a three-conductor cable; (b, c) equivalent circuits; (d) final capacitance circuit.

TABLE 3.2 Typical Values for Dielectric Constants of Cable Insulation

Type of Insulation	Permittivity (ε)
Polyvinyl chloride (PVC)	3.5–8.0
Ethylene–propylene insulation (EP)	2.8–3.5
Polyethylene insulation	2.3
Cross-linked polyethylene	2.3–6.0
Impregnated paper	3.3–3.7

This book mainly confines to power frequency models, except for some sporadic references to high-frequency models [28–35].

3.7 EMTP MODELS

EMTP permits a number of transmission line models for the transient studies. The ⊓ model can be used for steady-state or frequency scan solutions and is not valid for time-domain solutions. The model data produced is in terms of a Y-matrix representation that includes series and shunt branches.

The constant parameter (CP) model is a frequency independent transmission line model for the wave equation of the distributed parameter line. It is not accurate for zero sequence currents or high-frequency phenomena. It can be successfully used for problem analysis with limited frequency dispersion.

J. Marti's frequency dependent model [34] is more accurate than the CP model, though computationally slower. It takes in to account the frequency dependence of series resistance and inductance of the line. This model is also based upon modal decomposition techniques, not fully discussed in this book.

EMTP uses the following transformation matrices. The Clarke's $\alpha\beta0$ transformation matrix for m-phase balanced line is

$$
\bar{T}_i =
\begin{vmatrix}
\dfrac{1}{\sqrt{m}} & \dfrac{1}{\sqrt{2}} & \dfrac{1}{\sqrt{6}} & \cdot & \dfrac{1}{\sqrt{j(j-1)}} & \cdot & \dfrac{1}{\sqrt{m(m-1)}} \\[2ex]
\dfrac{1}{\sqrt{m}} & -\dfrac{1}{\sqrt{2}} & \dfrac{1}{\sqrt{6}} & \cdot & \dfrac{1}{\sqrt{j(j-1)}} & \cdot & \dfrac{1}{\sqrt{m(m-1)}} \\[2ex]
\dfrac{1}{\sqrt{m}} & 0 & -\dfrac{2}{\sqrt{6}} & \cdot & \cdot & \cdot & \cdot \\[2ex]
\cdot & \cdot & 0 & \cdot & \dfrac{-(j-1)}{\sqrt{j(j-1)}} & \cdot & \cdot \\[2ex]
\cdot & \cdot & \cdot & \cdot & 0 & \cdot & \cdot \\[2ex]
\dfrac{1}{\sqrt{m}} & 0 & 0 & \cdot & 0 & \cdot & \dfrac{-(m-1)}{\sqrt{m(m-1)}}
\end{vmatrix}
\tag{3.48}
$$

Applying this m-phase transformation to matrices of m-phase balanced lines will produce a diagonal matrix of the form:

$$\begin{vmatrix} Z_{g-m} & & \\ & Z_{L-m} & \\ & & Z_{L-m} \end{vmatrix} \qquad - \qquad (3.49)$$

Z_{g-m} is ground mode and Z_{l-m} are the line mode matrices. The solution becomes simpler if m-phase transmission line equations (M-coupled equations) can be transformed into M-decoupled equations. Many transposed and even un-transposed lines can be diagonalized with transformations to modal parameters based upon eigenvalue/eigenvector theory. The phase differential equation:

$$\frac{d^2 V_{ph}}{dx^2} = \bar{Z}_{ph} \bar{Y}_{ph} V_{ph} \qquad (3.50)$$

becomes

$$\frac{d^2 V_{mode}}{dx^2} = \bar{\Lambda} V_{mode} \qquad (3.51)$$

where

$$\bar{\Lambda} = T_v^{-1} \bar{Z}_{ph} \bar{Y}_{ph} \bar{T}_v \qquad (3.52)$$

The diagonal elements of $\bar{\Lambda}$ are eigenvalues of matrix product $\bar{Z}_{ph}\bar{Y}_{ph}$ and $\bar{T}_v$ is the matrix of eigenvectors or modal matrix of that matrix product. Some methods of finding eigenvectors and eigenvalues are Q-R transformation and iteration schemes.

Similarly for current,

$$\bar{I}_{\mathrm{mode}} = \bar{T}_i \bar{I}_{ph} \qquad (3.53)$$

$$\frac{d^2 V_{\mathrm{mode}}}{dx^2} = \bar{\Lambda} V_{\mathrm{mode}} \qquad (3.54)$$

$$\bar{T}_i = [\bar{T}_v^t]^{-1} \qquad (3.55)$$

For a three-phase line,

$$\bar{T}_i = \begin{vmatrix} \dfrac{1}{\sqrt{3}} & \dfrac{1}{\sqrt{2}} & \dfrac{1}{\sqrt{6}} \\[2mm] \dfrac{1}{\sqrt{3}} & -\dfrac{1}{\sqrt{2}} & \dfrac{1}{\sqrt{6}} \\[2mm] \dfrac{1}{\sqrt{3}} & 0 & -\dfrac{2}{\sqrt{6}} \end{vmatrix} \qquad (3.56)$$

We can write

$$\bar{Z}_{012} = \bar{T}_i^{-1} \bar{Z}_{abc} \bar{T}_i \qquad (3.57)$$

Then,

$$
\bar{Z}_{123} =
\begin{vmatrix}
\dfrac{1}{\sqrt{3}} & \dfrac{1}{\sqrt{2}} & \dfrac{1}{\sqrt{6}} \\[2mm]
\dfrac{1}{\sqrt{3}} & -\dfrac{1}{\sqrt{2}} & \dfrac{1}{\sqrt{6}} \\[2mm]
\dfrac{1}{\sqrt{3}} & 0 & -\dfrac{2}{\sqrt{6}}
\end{vmatrix}^{-1}
\begin{vmatrix}
Z & M & M \\
M & Z & M \\
M & M & Z
\end{vmatrix}
\begin{vmatrix}
\dfrac{1}{\sqrt{3}} & \dfrac{1}{\sqrt{2}} & \dfrac{1}{\sqrt{6}} \\[2mm]
\dfrac{1}{\sqrt{3}} & -\dfrac{1}{\sqrt{2}} & \dfrac{1}{\sqrt{6}} \\[2mm]
\dfrac{1}{\sqrt{3}} & 0 & -\dfrac{2}{\sqrt{6}}
\end{vmatrix}
$$

$$
=
\begin{vmatrix}
Z+2M & 0 & 0 \\
0 & Z-M & 0 \\
0 & 0 & Z-M
\end{vmatrix}
\tag{3.58}
$$

In Chapter 1, we had the same result using symmetrical components.

Similarly for the shunt elements,

$$
\hat{\bar{Y}} = \bar{T}_v^{-1}\bar{Y}\,\bar{T}_v
\tag{3.59}
$$

Table 3.1 qualifies representation of transposed lines. For this case modal decoupling is still possible, however transition matrices T_i and T_v are different for each line configuration and are a function of frequency.

3.7.1 Frequency Dependent Model, FD

The propagation constant can be defined as the ratio of the receiving end voltage to the source voltage for an open-ended line if the line is fed through its characteristics impedance (Figure 3.10 a). Then there will be no reflection from the far end. In this case $V_m + Z_c.I_{mk} = V_s$. We can write the receiving end voltage at k as

$$
\begin{aligned}
V_k &= V_s A(\omega) \\
\omega &= \exp(-\gamma l)
\end{aligned}
\tag{3.60}
$$

If a unit voltage from dc to all frequencies is applied at the source end, then its time response will be unit impulse, infinitely narrow (area $= 1.0$), and integral of voltage $=$ unit step. We can write time response, $a(t)$ to a unit impulse as inverse Fourier transform of $A(\omega)$. ($A(\omega) = e^{-\gamma l}$). This will not be attenuated and no longer infinitely narrow. The impulse response for a lossless line is unit impulse at $t = \tau$ with area 1.0. Setting $V_{source} = 1.0$ in Equation (3.60) means that $A(\omega)$ transformed into time domain must be an impulse; which arrives at the other end k, if the source is a unit impulse.

The history term V_m/Z_c+I_{mk} at t-τ is picked up and weighted with $a(t)$. This weighting at other end of line is done with convolution integral:

$$
hist_{propagation} = \int_{\tau=\min}^{\tau=\max} I_{m\text{-total}}(t-u)a(u)dt
\tag{3.61}
$$

which can be evaluated with recursive convolution. $I_{m\text{-total}}$ is the sum of the line current I_{mk} and a current which will flow through characteristic impedance if a voltage

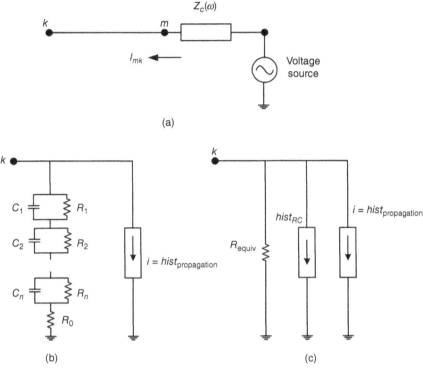

Figure 3.10   Explanation of J. Marti frequency dependent (FD) model of transmission line in EMTP; (a) voltage source connected through matching impedance to node m; (b) RC network; (c) circuit with equivalent resistance after applying implicit integration.

V_m is applied to it. The approximation of Z_c, a frequency dependent impedance, is done with a Foster-I-R-C network (Figure 3.10b). Applying trapezoidal rule of integration, each RC block is a current source in parallel with an equivalent resistance. Summing these up, a frequency dependent line is represented as shown in Figure 3.10c. J. Marti shows that it is best to sum up $A(\omega)$ and $Z(\omega)$ in the frequency domain.

$$A(s) = e^{-s\tau_{\min}} k \frac{(s + z_1)(s + z_2) \cdots \cdot (s + z_n)}{(s + p_1)(s + p_2) \cdots (s + p_n)} \qquad (3.62)$$

Partial fractions can be used and the time response is

$$a(t) = \left[k_1 e^{-p_1(t-\tau_{\min})} + k_2 e^{-p_2(t-\tau_{\min})} \dots + k_m e^{-p_m(t-\tau_{\min})} \right] \quad for \ t \geq \tau_{\min}$$
$$= 0 \quad for \ t < \tau_{\min}. \qquad (3.63)$$

The weighting factor is used to calculate the history term in each time step. Similar expression of a transfer function of poles and zeros for Z_c is applicable.

$$Z_c(s) = k \frac{(s + z_1)(s + z_2) \cdots \cdot (s + z_n)}{(s + p_1)(s + p_2) \cdots (s + p_n)} \qquad (3.64)$$

The success depends upon the quality of approximation for $A(\omega)$ and $Zc(\omega)$. J. Marti used Bode's procedure for approximating the magnitudes of the functions. The M-phase lines, any of the M-modes can be specified as frequency dependent or with lumped resistances or distortionless. Field tests have verified the accuracy of calculations [35].

3.8 EFFECT OF HARMONICS ON LINE MODELS

Long-line effects should be represented for lines of length $150/h$ miles, where h is the harmonic number. The effect of higher frequencies is to increase the skin effect and proximity effects. A frequency-dependent model of the resistive component becomes important, though the effect on the reactance is ignored. The resistance can be multiplied by a factor $g(h)$

$$R(h) = R_{dc}g(h) \tag{3.65}$$

$$g(h) = 0.035X^2 + 0.938 > 2.4 \tag{3.66}$$

$$= 0.35X + 0.3 \le 2.4 \tag{3.67}$$

where

$$X = 0.3884\sqrt{\frac{f_h}{f}}\sqrt{\frac{h}{R_{dc}}} \tag{3.68}$$

Where f_h = harmonic frequency and f = system frequency. Another equation for taking into account the skin effect is

$$R = R_e\left(\frac{j\mu\omega}{2\pi a}\frac{J_z(r)}{\partial J_z(r)/\partial r\,|_{r=a}}\right) \tag{3.69}$$

where J_z (r) is the current density, and a is the outside radius of the conductor.

3.9 TRANSMISSION LINE EQUATIONS WITH HARMONICS

In presence of harmonics, from Table 3.1,

$$V_{s(h)} = V_{r(h)}\cosh(\gamma l_{(h)}) + I_{r(h)}Z_{0(h)}\sinh(\gamma l_{(h)})$$

$$= V_{r(h)}\cosh\left(\sqrt{Z_{(h)}Y_{(h)}}\right) + I_{r(h)}\sqrt{\frac{z_{(h)}}{y_{(h)}}}\sinh\left(\sqrt{Z_{(h)}Y_{(h)}}\right) \tag{3.70}$$

$$I_{s(h)} = \frac{V_{r(h)}}{Z_{0(h)}}\sinh\left(\sqrt{Z_{(h)}Y_{(h)}}\right) + I_{r(h)}\cosh\left(\sqrt{Z_{(h)}Y_{(h)}}\right) \tag{3.71}$$

Similarly from

$$\begin{vmatrix} V_r \\ I_r \end{vmatrix} = \begin{vmatrix} \cosh(\gamma l) & -Z_0\sinh(\gamma l) \\ -\dfrac{\sinh(\gamma l)}{Z_0} & \cosh\gamma l \end{vmatrix} \begin{vmatrix} V_s \\ I_s \end{vmatrix} \qquad (3.72)$$

equations of receiving end current and voltages can be written.

The variation of the impedance of the transmission line with respect to frequency is of much interest. Table 3.1 shows the constants for a nominal Π model. An equivalent Π model, which relates the Π model with long line model, can be derived.

$$\begin{aligned} Z_s &= Z_0\sinh\gamma l \\ Y_p &= \frac{Y}{2}\left[\frac{\tanh\gamma l/2}{\gamma l/2}\right] \end{aligned} \qquad (3.73)$$

Here we have denoted the series element of the Π-model with Z_s and the shunt element with Y_p. In presence of harmonics, we can write

$$\begin{aligned} Z_{s(h)} &= Z_{0(h)}\sinh\gamma l_{(h)} = Z_{(h)}\frac{\sinh\gamma l_{(h)}}{\gamma l_{(h)}} \\ Y_{p(h)} &= \frac{\tanh\left(\gamma l_{(h)}/2\right)}{Z_{0(h)}} = \frac{Y_{(h)}}{2}\frac{\tanh\left(\gamma l_{(h)}/2\right)}{\gamma l_{(h)}/2} \end{aligned} \qquad (3.74)$$

and

$$\begin{aligned} \gamma_{(h)} &= \sqrt{z_{(h)}y_{(h)}} \approx \frac{h\omega}{l}\sqrt{LC} \\ Z_{0(h)} &= \sqrt{z_{(h)}/y_{(h)}} \approx \sqrt{\frac{L}{C}} \end{aligned} \qquad (3.75)$$

Also,

$$\begin{aligned} \lambda_{(h)} &= 2\pi/\beta_{(h)} \approx \frac{l}{hf\sqrt{LC}} \\ v_{(h)} &= f\lambda_h \approx \frac{l}{\sqrt{LC}} \\ f_{osc(h)} &= \frac{v_{(h)}}{l} \approx \frac{1}{\sqrt{LC}} \end{aligned} \qquad (3.76)$$

Thus, the characteristic impedance, velocity of propagation, and frequency of oscillations are all independent of h while wavelength varies inversely with h.

The impedance, resistance, and reactance plots of transmission lines versus frequency depend upon the line model used and there are much variations in these plots and resonant frequencies exhibited depending upon the line model.

We can also write the following equations:

$$
\begin{vmatrix} I_s \\ I_r \end{vmatrix} = \frac{1}{B} \begin{vmatrix} D & CB - DA \\ 1 & -A \end{vmatrix} \begin{vmatrix} V_s \\ V_r \end{vmatrix}
$$

$$
= \frac{1}{Z_0 \sinh \gamma l} \begin{vmatrix} \cosh \gamma l & -1 \\ 1 & -\cosh \gamma l \end{vmatrix} \begin{vmatrix} V_s \\ V_r \end{vmatrix} \tag{3.77}
$$

$$
= \begin{vmatrix} \dfrac{1}{Z_s} + Y_p & -\dfrac{1}{Z_s} \\ \dfrac{1}{Z_s} & -\dfrac{1}{Z_s} - Y_p \end{vmatrix} \begin{vmatrix} V_s \\ V_r \end{vmatrix}
$$

Note that

$$
CB - DA = \sinh^2(\gamma l) - \cosh^2(\gamma l) = -1
$$

Looking from one end, the impedance of the line is

$$
\frac{1}{Y_{p(h)}} \text{ in parallel with } \left(Z_{s(h)} + \frac{1}{Y_{p(h)}} \right)
$$

$$
= \frac{Z_{s(h)} Y_{p(h)} + 1}{Y_{p(h)} \left[Z_{s(h)} Y_{p(h)} + 2 \right]}. \tag{3.78}
$$

A computer program is required to calculate the impedance along the line using Equation (3.78) with incremental changes in the frequency to locate the resonance points.

Example 3.5   Consider a transmission line with the following parameters: conductor "FLINT", 740.8 KCMIL, ACSR, 37 strands, spaced horizontally 25 ft apart, one ground wire placed 16 ft above the conductors, height of conductors from ground = 60 ft, soil resistivity = 90 Ω-m.
A computer program is used to calculate the line parameters:

$$
R = 0.14852 \, \Omega/\text{mile}
$$
$$
X = 0.837 \, \Omega/\text{mile}
$$
$$
Y = 5.15 \times 10^{-6} \, \text{S/mile}
$$

Then

$L = 0.00222 \, \text{H/mile}$

$C = 0.01366 \times 10^{-6} \, \text{F/mile}$

$Z_0 = 403 \, \Omega$

$\gamma = \sqrt{zy} = [(0.14852 + j0.837)(j5.15 \times 10^{-6}]^{1/2} = (0.184 + j2.084) \times 10^{-3}/\text{mile}$

$\lambda = \dfrac{2\pi}{\beta} = 3014 \, \text{mile}$

$v = f\lambda = 1.808 \times 10^5 \, \text{mile/s}$

$f_{osc} = \dfrac{1}{\sqrt{LC}} = 3016 \, \text{Hz (300 mile-line)}$

Also:

$$Z_s = Z_c \sinh(\gamma l) = \sqrt{\frac{0.14852 + j0.837}{j5.15 \times 10^{-6}}} \sinh(0.0552 + j0.6252)$$

$$= (405 - j36)\sinh(0.0552 + j0.6252)$$

$$\sinh(0.0552 + j0.6252) = \sinh(0.0552)\cos(0.6252) + j\cosh(0.0552)\sin(0.6252)$$

$$= (0.0552)(0.8108) + j(1.0)(0.5853)$$
$$= 0.045 + j0.5853$$

Then

$$Z_s = (405 - j36)(0.045 + j0.5833)$$
$$= 39.224 + j234.6$$
$$Y_p = \tanh(\gamma l/2)/Z_c$$
$$\tanh(\gamma l/2) = \frac{\sinh(\gamma l/2)}{\cosh(\gamma l/2)}$$

$$\sinh(\gamma l/2) = \sinh(0.0275 + j0.326) = (0.0275)(0.947) + j(1)(0.32) = 0.026 + j0.32$$
$$\cosh(\gamma l/2) = \cosh(0.275)(0.947) + j(0.0275)(0.32) = 0.947 + j0.0088$$
$$\tanh(\gamma l/2) = 0.031 + j0.338$$
$$Y_p = \frac{0.031 + j0.338}{405 - j36} = 2.341 \times 10^{-6} + j8.348 \times 10^{-4} \approx j8.348 \times 10^{-4}$$

Then the line impedance at fundamental frequency is

$$\frac{Z_{s(h)}Y_{p(h)} + 1}{Y_{p(h)}\left[Z_{s(h)}Y_{p(h)} + 2\right]}$$

$$= \frac{(405 - j36)(j8.348 \times 10^{-4}) + 1}{(j8.348 \times 10^{-4})[(405 - j36)(j8.348 \times 10^{-4}) + 2]}$$

$$= \frac{1.03 + j0.338}{-2.822 \times 10^{-4} + 1.695 \times 10^{-3})}$$

$$= 98.46 - j600.36$$

This shows the procedure. A computer program will run these calculations at close increment of frequency to capture the series and shunt resonance frequencies; plots as demonstrated in Figures 3.11 a and 3.11b.

Figure 3.11a shows the calculated impedance angle and Figure 3.12(b) shows the impedance modulus versus frequency plots of a 400 kV line, consisting of four bundled conductors of 397.5 kcmil ACSR per phase in horizontal configuration, GMD = 44.09', 200 miles long at no load: looking from the sending end. The line generates 194.8 Mvar of capacitive reactive power. When the line is loaded to 200 MVA at 0.9 PF (no harmonics), the impedance angle and modulus (which is much reduced) are shown in Figures 3.11c and 3.11d. Two ground wires of 7#8 are considered in the calculations, and the earth resistivity is 100 Ω m. Each conductor has a diameter of 0.806 in. (0.317 cm) and bundle conductors are spaced 6 in. (0.15 m) center to center. The plots show a number of resonant frequencies. The impedance angle changes

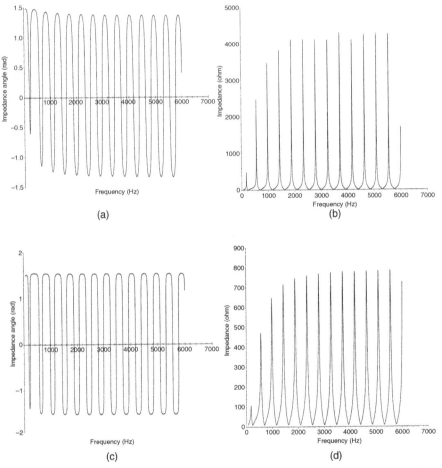

Figure 3.11 (a) frequency scan of a 400 kV line with bundle conductors; (b) corresponding phase angle of the impedance modulus; (c, d) corresponding plots with line loaded to 200 MVA, 0.9 lagging power factor. There are no harmonic generating loads, see text.

abruptly at each resonant frequency. For harmonic analyses of HV long transmission lines, a computer program which can model the long line model with harmonics is required.

REFERENCES

[1] Central Station Engineers. *Electrical Transmission and Distribution Reference Book*, 4th edition. East Pittsburgh, PA: Westinghouse Corp., 1964.

[2] *Transmission Line Reference Book—345 kV and Above*. Palo Alto, CA: EPRI, 1975.

[3] Special Issue on EHV Transmission, *IEEE Transactions on Power Apparatus and Systems*, vol. PAS-85, no. 6, pp. 555–700, 1966.

[4] DG Fink (ed.). *Standard Handbook for Electrical Engineers*, 10th edition. New York: McGraw-Hill, 1969.

[5] T Croft, JH Watt, and C Carr. *American Electrician's Handbook*, 9th edition. New York: McGraw-Hill, 1970.

[6] T Gonen. *Electrical Power Transmission System Engineering: Analysis and Design*, 2nd edition, Boca Raton, FL: CRC Press, 2009.

[7] The Aluminum Association. *Aluminum Conductor Handbook*, 2nd edition. Washington, DC, 1982.

[8] WD Stevenson. *Elements of Power System Analysis*, 2nd edition. McGraw-Hill, 1962.

[9] A Semlyen. Overhead line parameters form handbook formulas and computer programs. *IEEE Transactions on Power and Systems*, vol. PAS-104, p. 371, February 1985.

[10] P Chowdhri. *Electromagnetic Transients in Power Systems*. John Wiley & Sons, 1996.

[11] LV Beweley. *Traveling Waves on Transmission Systems*, 2nd edition. New York: John Wiley& Sons, 1951.

[12] R Rudenberg. *Transient Performance of Electrical Systems*. New York: McGraw-Hill, 1950.

[13] MS Naidu and V Kamaraju. *High Voltage Engineering*, 2nd edition. New Delhi: McGraw-Hill, 1999.

[14] G Anderson. *Transmission Reference Book*. New York: Edison Electric Company, 1968.

[15] *Westinghouse Electric Transmission and Distribution Handbook*, 4th edition. East Pittsburgh, PA: Westinghouse Electric Corp., 1964.

[16] PM Anderson. *Analysis of Faulted Systems*. Ames, IA: Iowa State University Press, 1973.

[17] JC Das. *Power System Analysis: Short-Circuit Load Flow and Harmonics*, 2nd edition. Boca Raton, FL: CRC Press, 2012.

[18] HW Beaty and DG Fink (eds). *Standard Handbook for Electrical Engineers*, 15th edition. New York: McGraw-Hill, 2007.

[19] G Bianchi and G Luoni. Induced currents and losses in single-core submarine cables. *IEEE Transactions on Power Apparatus and Systems*, vol. PAS-95, pp. 49–58, February 1976.

[20] A Ammentani. A general formation of impedance and admittance of cables. *IEEE Transactions on Power and Systems,* vol. PAS-99, pp. 902–910, May/June 1980.

[21] DR Smith and JV Bager. Impedance and circulating current calculations for UD multi-wire concentric neutral circuits. *IEEE Transactions on Power and Systems*, vol. PAS-91, pp. 992–1006, May/June 1972.

[22] P Arizon and HW Dommel. Computation of cable impedances based on subdivision of conductors. *IEEE Transactions on Power Delivery,* vol. PWRD-2, pp. 21–27, January 1987.

[23] IEEE Std. 525, IEEE Guide for the Design and Installation of Cable Systems in Substations, 1987.

[24] LM Wedepohl and DJ Wilcox. Transient analysis of underground power transmission systems: system model and wave propagation characteristics. *Proceedings of the IEEE*, vol. 120, pp. 252–259, February 1973.

[25] GW Brown and RG Rocamora. Surge propagation in three-phase pipe-type cables, Part I: Unsaturated pipe. *IEEE Transactions on Power Apparatus and Systems,* vol. PAS-90, pp. 1287–1294, May 1971.

[26] GW Brown and RG Rocamora. Surge propagation in three-phase pipe-type cables, Part II: Duplication of field tests including the effect of neutral wires and pipe saturation. *IEEE Transactions on Power Apparatus and Systems*, vol. 96, no. 3, pp. 826–833, May/June 1977.

[27] IEEE P575/D12, Draft Guide for Bonding Shields and Sheaths of Single-Conductor Power Cables Rated 5 kV through 500 kV, 2013.

[28] L Marti. Simulation of electromagnetic transients in underground cables with frequency-dependent model transformation matrices. *IEEE Transactions on Power Delivery*, vol. 3, no. 3, pp. 1099–1110, 1988.

[29] CIGRE, Working Group 33.02 (Internal Voltages), Guide Lines for Representation of Network Elements when Calculating Transients, Paris, 1990.

[30] JC Das. *Transients in Electrical Systems*. New York: McGraw-Hill, 2010.

[31] Canadian/American EMTP User Group. *ATP Rule Book*. Portland: Oregon, 1987–1992.

[32] IEEE Task Force Report. Review of technical considerations on limits of interference from power lines and stations. *IEEE Transactions on Power Apparatus and Systems,* vol. PAS-99, no.1, p. 365, January/February 1980.

[33] LM Wedepohl and DJ Wilcox. Transient analysis of underground power transmission systems: system model and wave propagation characteristics. *Proceedings of the IEEE*, vol. 120, pp. 252–259, February 1973.

[34] JR Marti. The problem of frequency dependence in transmission line modeling. PhD thesis, The University of British Columbia, Vancouver, Canada, April 1981.

[35] WS Meyer and HW Dommel. Numerical modeling of frequency dependent transmission line parameters in EMTP. *IEEE Transactions on Power Apparatus and Systems*, vol. PAS93, pp. 1401–1409, September /October 1974.

CHAPTER **4**

SEQUENCE IMPEDANCES OF ROTATING EQUIPMENT AND STATIC LOAD

4.1 SYNCHRONOUS GENERATORS

Table 4.1 shows the data for a 112.1 MVA, 2-pole synchronous generator. This book does not provide an explanation or definitions of the parameters shown in this Table. References [1–4] may be seen.

4.1.1 Positive Sequence Impedance

The positive sequence impedance is not visible in Table 4.1. Depending upon the type of study, the positive sequence impedance may be considered as;

- X_{dv}'' for the short-circuit calculations according to ANSI/IEEE standards. The saturated value of the subtransient reactance is used.
- X_d'' for the subtransient conditions.
- X_d' for the transient conditions.
- X_d for the steady-state conditions.

The subtransient, transient and steady-state conditions can be defined with respect to terminal three-phase short-circuit of the synchronous generator [4]. Each of the parameters in Table 4.1 can be defined with specific equations, not shown here.

The bullet points 2, 3, and 4 ignore *saliency.*

The positive sequence resistance is the ac resistance of the armature windings.

Saturation varies with voltage, current, and power factor. The saturation factor is usually applied to transient and synchronous reactances, though all other reactances change, though slightly, with saturation. In a typical machine, transient reactances may reduce from 5% to 25% on saturation. Saturated reactance is sometimes called the rated voltage reactance and is denoted by subscript "v" added to the "d" and "q" subscript axes, that is, X_{dv}'' and X_{qv} denote saturated subtransient reactances in direct and quadrature axes, respectively. For calculations of short-circuit currents according to ANSI/IEEE standards saturated subtransient reactance must be used.

Understanding Symmetrical Components for Power System Modeling, First Edition. J.C. Das.
© 2017 by The Institute of Electrical and Electronics Engineers, Inc. Published 2017 by John Wiley & Sons, Inc.

TABLE 4.1 Generator Data

Description	Symbol	Data
Generator		
112.1 MVA, 2-pole, 13.8 kV, 0.85 pf, 95.285 MW, 4690 A, SCR, 235 field V, wye connected	0.56	
Per unit reactance data, direct axis		
Saturated synchronous	X_{dv}	1.949
Unsaturated synchronous	X_d	1.949
Saturated transient	X'_{dv}	0.207
Unsaturated transient	X'_d	0.278
Saturated subtransient	X''_{dv}	0.164
Unsaturated subtransient	X''_d	0.193
Saturated negative sequence	X_{2v}	0.137
Unsaturated negative sequence	X_{2I}	0.185
Saturated zero sequence	X_{0v}	0.092
Leakage reactance, overexcited	X_{0I}	0.111
Leakage reactance, underexcited	$X_{LM,OXE}$	0.164
	$X_{LM,UEX}$	0.164
Per unit reactance data, quadrature axis		
Saturated synchronous	X_{qv}	1.858
Unsaturated synchronous	X_q	1.858
Unsaturated transient	X'_q	0.434
Saturated subtransient	X''_{qv}	0.140
Unsaturated subtransient	X''_q	0.192
Field time constant data, direct axis		
Open circuit	T'_{d0}	5.615
Three-phase short-circuit transient	T'_{d3}	0.597
Line-to-line short-circuit transient	T'_{d2}	0.927
Line-to-neutral short-circuit transient	T'_{d1}	1.124
Short-circuit subtransient	T''_d	0.015
Open circuit subtransient	T''_{d0}	0.022
Field time constant data quadrature axis		
Open circuit	T'_{q0}	0.451
Three-phase short-circuit transient	T'_q	0.451
Short-circuit subtransient	T''_q	0.015
Open circuit subtransient	T''_{q0}	0.046
Armature dc component time constant data		
Three-phase short-circuit	T_{a3}	0.330
Line-to-line short-circuit	T_{a2}	0.330
Line-to-neutral short-circuit	T_{a1}	0.294

4.1.2 Negative Sequence Impedance

If we apply reverse phase sequence currents (negative sequence currents) to the armature winding, and the generator is running at synchronous speed, the field winding shorted through exciter, then the ratio of the negative sequence fundamental frequency voltages to the currents gives the negative sequence impedance. The field flux due to negative sequence currents rotates at synchronous speed opposite to the direction of the rotor. As viewed from stator, the stator currents appear to be double frequency currents in i_d, and i_q, that is in direct and quadrature axes. It is shown in many texts that the negative sequence reactance can be written as

$$X_2 = \frac{X_d'' + X_q''}{2} \tag{4.1}$$

The negative sequence flux component in the air gap may be considered to alternate between poles and interpolar gap, respectively.

The power associated with negative sequence current can be expressed as resistance multiplied by the square of the current, and this resistance is termed as negative sequence resistance.

4.1.3 Negative Sequence Capability of Generators

Synchronous generators have limited capability to withstand negative sequence currents. According to NEMA [5] specifications, a synchronous machine shall be capable of withstanding, without injury, a 30-s, three-phase short-circuit at its terminals when operating at rated kVA and power factor, at 5% overvoltage, with fixed excitation. With a voltage regulator in service, the allowable duration t, in seconds, is determined from the following equation, where the regulator is designed to provide a ceiling voltage continuously during a short-circuit:

$$t = \left(\frac{\text{Norminal field voltage}}{\text{Exciter ceiling voltage}}\right)^2 \times 30\,\text{s} \tag{4.2}$$

The generator should also be capable of withstanding, without injury, *any other* short-circuit at its terminals for 30 s, provided that

$$I_2^2 t \leq 40 \text{ for salient-pole machines} \tag{4.3}$$

$$I_2^2 t \leq 30 \text{ for air-cooled cylindrical rotor machines} \tag{4.4}$$

and the maximum current is limited by external means so as not to exceed the three-phase fault; I_2 is the negative sequence current due to unsymmetrical faults.

Synchronous generators have both a continuous and short-time unbalanced current capabilities, which are shown in Tables 4.2 and 4.3 [6, 7]. These capabilities are based upon 120 Hz negative sequence currents induced in the rotor due to continuous unbalance or unbalance under fault conditions. In absence of harmonics and impedance asymmetries (i.e., non transposition of transmission lines) it is a standard requirement that the synchronous generators should be able to supply some unbalance currents.

4.1.3.1 Effect of Harmonics
In a synchronous machine, the frequency induced in the rotor is net rotational difference between fundamental frequency and the harmonic frequency. Fifth harmonic rotates in reverse sequence with respect to stator and with respect to the rotor the

TABLE 4.2   Requirements of Unbalanced Faults on Synchronous Machines

Type of Synchronous Machine	Permissible $I_2^2 t$
Salient-pole generators	40
Synchronous condensers	30
Cylindrical rotor generators	
Indirectly cooled	30
Directly cooled (0–800 MVA)	10
Directly cooled (801–1600 MVA)	10-(00625)(MVA-800)^a

^{a}Thus for a 1600 MVA generator $I_2^2 t = 5$.

TABLE 4.3   Continuous Unbalance Current Capability of Generators

Type of Generator and Rating	Permissible I_2
Salient pole with connected amortisseur windings	10
Salient pole with non-connected amortisseur windings	5
Cylindrical rotor, indirectly cooled	10
Cylindrical rotor, directly cooled to	
960 MVA	8
961–1200 MVA	6
1201–1500 MVA	5

induced frequency is that of sixth harmonic. Similarly, the forward rotating 7th harmonic with respect to stator produces 6th harmonic in the rotor. The interaction of these fields produces a pulsating torque at 360 Hz and results in the oscillations of the shaft. Similarly, the harmonic pair 11 and 13 produces a rotor harmonic of 12th. If the frequency of the mechanical resonance exists close to these harmonics during starting, large mechanical forces can occur. The zero sequence harmonics ($h = 3$, 6, ...) do not produce a net flux density. These produce ohmic losses.

When the negative sequence capabilities are exploited for harmonic loading, the variations in loss intensity at different harmonics versus 120 Hz should be considered. The following expression can be used for equivalent heating effects of harmonics translated into negative sequence currents:

$$I_{2,\text{equiv}} = \left[\left(\frac{6f}{120} \right)^{1/2} (K_{5,7})(I_5 + I_7)^2 + \left(\frac{12f}{120} \right)^{1/2} (K_{11}, K_{13})(I_{11} + I_{13})^2 + \cdots \right]^{1/2}$$

(4.5)

where $K_{5,7}$, $K_{11,13}$, ... are correction factors to convert from maximum rotor surface loss intensity to average loss intensity [8] and these can be read from Figure 4.1, f = fundamental frequency and I_5, I_7 = harmonic current in pu values.

Example 4.1

    a. Consider a synchronous generator, with continuous unbalance capability of 0.10 pu, Table 4.3. It is subjected to 5th and 7th harmonic loading of 0.07 and 0.06 pu, respectively. Is the unbalance capability exceeded?

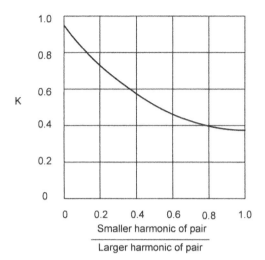

Figure 4.1   Ratio K for average loss to maximum loss based on harmonic pair.

From Figure 4.1 and harmonic ratio $0.06 / 0.07 = 0.857$, $K_{5,7} = 0.4$. From Equation (4.5),

$$I_{2,equiv} = \left[\sqrt{3}(0.4)(0.07 + 0.06)^2 \right]^{1/2} = 0.108$$

The continuous negative sequence capability is exceeded. If we simply sum up the harmonics we get 13%. Thus, calculation with simple summation is inaccurate.

b. A harmonic analysis study shows that a generator of 13.8 kV, 100 MVA, absorbs the following harmonic currents in percentage of the generator full load current: 5th harmonic = 40%, 7th harmonic = 30%, 11th harmonic = 18%, 13th harmonic = 10%, 17th harmonic = 5%, and 19th harmonic = 2%. The generator has a $I_2^2 t = 30$. The negative sequence relay for protecting the generator is set at 25 for conservatism. Find the operating time of the relay.

From Table 4.4, only the negative sequence harmonics need to be considered:

$$\left[(0.4)^2 + (0.18)^2 + (0.05)^2 \right] t = 25$$

The relay will trip in 128.3 s. The generator capability to withstand the negative sequence harmonics is 153 s. Harmonic loading of generators must be carefully calculated and synchronous generators must be protected for it.

4.1.4 Zero Sequence Impedance

The zero sequence impedance of generators with ungrounded neutrals has no significance as zero sequence currents cannot flow. It should, however, be noted that practically the generator neutrals are not left ungrounded, as damaging overvoltages can occur under faults.

TABLE 4.4 Harmonic Order and Sequence

Harmonic Order	Forward Positive	Reverse Negative	Zero
Fundamental	x		
2		x	
3			x
4	x		
5		x	
6			x
7	x		
8		x	
9			x
10	x		
11		x	x
12			
13	x		
14		x	
15			x
16	x		
17		x	
18			x
19	x		
20		x	
21			x
22	x		
23		x	
24			x
25	x		
26		x	
27			x
28	x		
29		x	
30			x

With zero sequence currents applied to generator armature windings, the instantaneous values of currents in three phases are equal. The net MMF produced is zero if we consider perfectly distributed and symmetrical windings. In a practical machine, the distribution is not sinusoidal and the flux due to three phases has a net small value, depending upon the winding construction, pitch, cording, breadth factors, etc.

Figure 4.2a shows the flow of zero sequence currents in three phases and neutral of a synchronous machine:

$$I_n = I_{a0} + I_{b0} + I_{c0} = 3I_{a0} \tag{4.6}$$

Thus, the equivalent circuit is shown in Figure 4.2b.

The zero sequence resistance is slightly larger than the positive sequence resistance.

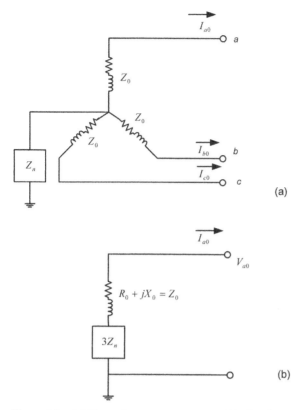

Figure 4.2   (a) Flow of zero sequence current in the phases and neutral of a synchronous generator; (b) equivalent circuit.

4.1.5 Sequence Component Transformation

The following expressions can be written for the terminal voltages of a wye connected synchronous generator, neutral grounded through an impedance Z_n (Figure 4.3).

$$V_a = \frac{d}{dt}\left[L_{af}\cos\theta I_f - L_{aa}I_a - L_{ab}I_b - L_{ac}I_c\right] - I_a R_a + V_n$$

$$V_b = \frac{d}{dt}\left[L_{bf}\cos(\theta - 120°)I_f - L_{ba}I_a - L_{bb}I_b - L_{bc}I_c\right] - I_a R_b + V_n \quad (4.7)$$

$$V_c = \frac{d}{dt}\left[L_{cf}\cos(\theta - 240°)I_f - L_{ca}I_a - L_{cb}I_b - L_{cc}I_c\right] - I_a R_c + V_n$$

The first term is the generator internal voltage, due to field linkages, and L_{af} denotes the field inductance with respect to phase A of stator windings and I_f is the field current. These internal voltages are displaced by 120°, and may be termed E_a, E_b, and E_c. The voltages due to armature reaction, given by the self-inductance of a phase, that is, L_{aa}, and its mutual inductance with respect to other phases, that is, L_{ab} and L_{ac}, and the IR_a drop are subtracted from the generator internal voltage and the neutral voltage is added to obtain the line terminal voltage V_a.

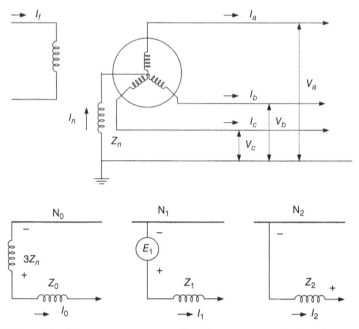

Figure 4.3   Sequence components of synchronous generator impedances.

For a symmetrical machine,

$$
\begin{aligned}
L_{af} &= L_{bf} = L_{cf} = L_f \\
R_a &= R_b = R_c = R \\
L_{aa} &= L_{bb} = L_{cc} = L \\
L_{ab} &= L_{bc} = L_{ca} = L'
\end{aligned}
\tag{4.8}
$$

Thus,

$$
\begin{vmatrix} V_a \\ V_b \\ V_c \end{vmatrix} = \begin{vmatrix} E_a \\ E_b \\ E_c \end{vmatrix} - j\omega \begin{vmatrix} L & L' & L' \\ L' & L & L' \\ L' & L' & L \end{vmatrix} \begin{vmatrix} I_a \\ I_b \\ I_c \end{vmatrix} - \begin{vmatrix} R & 0 & 0 \\ 0 & R & 0 \\ 0 & 0 & R \end{vmatrix} \begin{vmatrix} I_a \\ I_b \\ I_c \end{vmatrix} - Z_n \begin{vmatrix} I_n \\ I_n \\ I_n \end{vmatrix}
\tag{4.9}
$$

Transform using symmetrical components:

$$
T_s \begin{vmatrix} V_0 \\ V_1 \\ V_2 \end{vmatrix} = T_s \begin{vmatrix} E_0 \\ E_1 \\ E_2 \end{vmatrix} - j\omega \begin{vmatrix} L & L' & L' \\ L' & L & L' \\ L' & L' & L \end{vmatrix} T_s \begin{vmatrix} I_0 \\ I_1 \\ I_2 \end{vmatrix} - \begin{vmatrix} R & 0 & 0 \\ 0 & R & 0 \\ 0 & 0 & R \end{vmatrix} T_s \begin{vmatrix} I_0 \\ I_1 \\ I_2 \end{vmatrix} - 3Z_n \begin{vmatrix} I_0 \\ I_0 \\ I_0 \end{vmatrix}
$$

$$
\begin{vmatrix} V_0 \\ V_1 \\ V_2 \end{vmatrix} = \begin{vmatrix} E_0 \\ E_1 \\ E_2 \end{vmatrix} - j\omega \begin{vmatrix} L_0 & 0 & 0 \\ 0 & L_1 & 0 \\ 0 & 0 & L_2 \end{vmatrix} \begin{vmatrix} I_0 \\ I_1 \\ I_2 \end{vmatrix} - \begin{vmatrix} R & 0 & 0 \\ 0 & R & 0 \\ 0 & 0 & R \end{vmatrix} \begin{vmatrix} I_0 \\ I_1 \\ I_2 \end{vmatrix} - \begin{vmatrix} 3I_0 Z_n \\ 0 \\ 0 \end{vmatrix}
\tag{4.10}
$$

where

$$L_0 = L + 2L'$$
$$L_1 = L_2 + L - L' \tag{4.11}$$

The equation may, thus, be written as

$$\begin{vmatrix} V_0 \\ V_1 \\ V_2 \end{vmatrix} = \begin{vmatrix} 0 \\ E_1 \\ 0 \end{vmatrix} - \begin{vmatrix} Z_0 + 3Z_n & 0 & 0 \\ 0 & Z_1 & 0 \\ 0 & 0 & Z_2 \end{vmatrix} \begin{vmatrix} I_0 \\ I_1 \\ I_2 \end{vmatrix} \tag{4.12}$$

The equivalent circuits are shown in Figure 4.3. Even for a cylindrical rotor machine, the assumption $Z_1 = Z_2$ is not strictly valid. The resulting generator impedance matrix is nonsymmetrical.

4.1.6 Three-Phase Short-Circuit of a Generator

When the damper winding circuit is considered, the short-circuit current of a synchronous generator can be expressed as

$$i_a = \sqrt{2}E \left[\left(\frac{1}{X_d} \right) \sin(\omega t + \delta) + \left(\frac{1}{X'_d} - \frac{1}{X_d} \right) e^{-t/T'_d} \sin(\omega t + \delta) \right.$$

$$+ \left(\frac{1}{X''_d} - \frac{1}{X'_d} \right) e^{-t/T''_d} \sin(\omega t + \delta) \tag{4.13}$$

$$\left. - \frac{(X''_d + X''_q)}{2X''_d X''_q} e^{-t/T_a} \sin \delta - t \frac{(X''_d - X''_q)}{2X''_d X''_q} e^{-t/T_a} \sin(2\omega t + \delta) \right]$$

- The first term is final steady-state short-circuit current.
- The second term is normal-frequency decaying transient current.
- The third term is normal-frequency decaying subtransient current.
- The fourth term is asymmetric decaying dc current.
- The fifth term is double-frequency decaying current.

Figure 4.4 shows the decaying ac component only in the subtransient, transient, and steady-state region. The dc component makes the short-circuit current asymmetrical about the zero axes (Figure 4.5).

The proof is not shown here. It requires knowledge of Park's Transformation, basic concepts shown in Section 4.1.7.

Example 4.2   Calculate the component short-circuits currents at the instant of three-phase terminal short-circuit of the generator (particulars as shown in Table 4.1). Assume that phase a is aligned with the field at the instant of short-circuit, maximum asymmetry, that is, $\delta = 0$. The generator is operating at no-load prior to short-circuit.

The calculations are performed by substituting the required numerical data from Table 4.1 into Equation (4.13):

Steady-state current = 2.41 kA rms

Decaying transient current = 20.24 kA rms

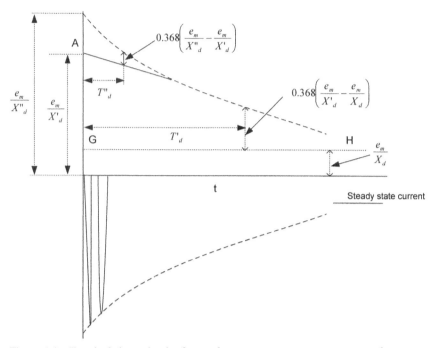

Figure 4.4   Terminal short-circuit of a synchronous generator, ac component decay; subtransient, transient, and steady-state currents.

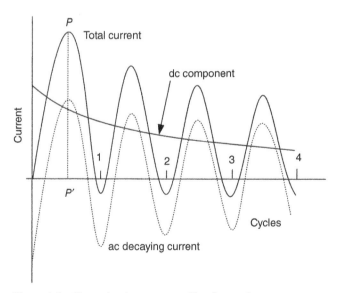

Figure 4.5   Short-circuit current profile of a synchronous generator with decaying ac and dc components.

Decaying subtransient current $= 5.95$ kA rms

Decaying dc component $= 43.95$ kA

Decaying second-harmonic component $= 2.35$ kA rms

Note that the second-harmonic component is zero if the direct axis and quadrature axis subtransient reactances are equal.

4.1.7 Park's Transformation

References [9, 10] are of great importance for the mathematical model of the synchronous machines for the transient stability analysis. They describe a new set of variables, such as currents, voltages, and flux linkages, obtained by transformation of the actual (stator) variables in three axes: 0, d, and q. Consider the normal construction of a three-phase synchronous machine, with three-phase stationary ac windings on the stator, and the field and damper windings on the rotor. *The stator inductances vary, depending on the relative position of the stator and rotor.*

The inductance matrix of a synchronous generator can be written as

$$\bar{L} = \begin{vmatrix} \bar{L}_{aa} & \bar{L}_{aR} \\ \bar{L}_{Ra} & \bar{L}_{RR} \end{vmatrix} \tag{4.14}$$

where $\bar{L}_{aa}$ is a stator-to-stator inductance matrix:

$$\bar{L}_{aa} = $$
$$\begin{vmatrix} L_s + L_m \cos 2\theta & -M_s - L_m \cos 2\left(\theta + \pi/6\right) & -M_s - L_m \cos 2\left(\theta + 5\pi/6\right) \\ -M_s - L_m \cos 2\left(\theta + \pi/6\right) & L_s + L_m \cos 2\left(\theta - 2\pi/3\right) & -M_s - L_m \cos 2\left(\theta - \pi/2\right) \\ -M_s - L_m \cos 2\left(\theta + 5\pi/6\right) & -M_s - L_m \cos 2\left(\theta - \pi/2\right) & L_s + L_m \cos 2\left(\theta + 2\pi/3\right) \end{vmatrix} \tag{4.15}$$

$\bar{L}_{aR} = \bar{L}_{Ra}$ is the stator-to-rotor inductance matrix:

$$\bar{L}_{aR} = \bar{L}_{Ra} = \begin{vmatrix} M_F \cos \theta & M_D \cos \theta & M_Q \sin \theta \\ M_F \cos(\theta - 2\pi/3) & M_D \cos(\theta - 2\pi/3) & M_Q \sin(\theta - 2\pi/3) \\ M_F \cos(\theta + 2\pi/3) & M_D \cos(\theta + 2\pi/3) & M_Q \sin(\theta + 2\pi/3) \end{vmatrix} \tag{4.16}$$

$\bar{L}_{RR}$ is the rotor-to-rotor inductance matrix:

$$\bar{L}_{RR} = \begin{vmatrix} L_F & M_R & 0 \\ M_R & L_D & 0 \\ 0 & 0 & L_Q \end{vmatrix} \tag{4.17}$$

The inductance matrix of Equation (4.15) shows that the inductances vary with the angle θ. By referring the stator quantities to rotating rotor dq axes through Park's transformation, this dependence on θ is removed and a constant reactance matrix emerges.

Park's transformation describes a new set of variables such as currents, voltages, and flux linkages in $0dq$ axes. The stator parameters are transferred to the rotor parameters. For the currents this transformation is

$$\begin{vmatrix} i_0 \\ i_d \\ i_q \end{vmatrix} = \sqrt{\frac{2}{3}} \begin{vmatrix} \dfrac{1}{\sqrt{2}} & \dfrac{1}{\sqrt{2}} & \dfrac{1}{\sqrt{2}} \\ \cos\theta & \cos\left(\theta - 2\dfrac{\pi}{3}\right) & \cos\left(\theta + 2\dfrac{\pi}{3}\right) \\ \sin\theta & \sin\left(\theta - 2\dfrac{\pi}{3}\right) & \sin\left(\theta + 2\dfrac{\pi}{3}\right) \end{vmatrix} \begin{vmatrix} i_a \\ i_b \\ i_c \end{vmatrix} \tag{4.18}$$

Using matrix notation,

$$\bar{i}_{0dq} = \bar{P}\bar{i}_{abc} \tag{4.19}$$

Similarly,

$$\bar{v}_{0dq} = \bar{P}\bar{v}_{abc} \tag{4.20}$$

$$\bar{\lambda}_{0dq} = \bar{P}\bar{\lambda}_{abc} \tag{4.21}$$

where $\bar{\lambda}$ is the flux linkage vector. The a–b–c currents in the stator windings produce a synchronously rotating field, stationary with respect to the rotor. This rotating field can be produced by constant currents in the fictitious rotating coils in the dq axes; P is nonsingular and $\bar{P}^{-1} = \bar{P}'$.

$$\bar{P}^{-1} = \bar{P}' = \sqrt{\frac{2}{3}} \begin{vmatrix} \dfrac{1}{\sqrt{2}} & \cos\theta & \sin\theta \\ \dfrac{1}{\sqrt{2}} & \cos\left(\theta - \dfrac{2\pi}{3}\right) & \sin\left(\theta - \dfrac{2\pi}{3}\right) \\ \dfrac{1}{\sqrt{2}} & \cos\left(\theta + \dfrac{2\pi}{3}\right) & \sin\left(\theta + \dfrac{2\pi}{3}\right) \end{vmatrix} \tag{4.22}$$

To transform the stator-based variables into rotor-based variables, define a matrix as follows:

$$\begin{vmatrix} i_0 \\ i_d \\ i_q \\ i_F \\ i_D \\ i_Q \end{vmatrix} = \begin{vmatrix} \bar{P} & \bar{0} \\ \bar{0} & \bar{1} \end{vmatrix} \begin{vmatrix} i_a \\ i_b \\ i_c \\ i_F \\ i_D \\ i_Q \end{vmatrix} = \bar{B}\bar{i} \tag{4.23}$$

where $\bar{1}$ is a 3×3 unity matrix and $\bar{0}$ is a 3×3 zero matrix. The original rotor quantities are left unchanged. The time-varying inductances can be simplified by referring all quantities to the rotor frame of reference:

$$\begin{vmatrix} \bar{\lambda}_{0dq} \\ \bar{\lambda}_{FDQ} \end{vmatrix} = \begin{vmatrix} \bar{P} & \bar{0} \\ \bar{0} & \bar{1} \end{vmatrix} \begin{vmatrix} \bar{\lambda}_{abc} \\ \bar{\lambda}_{FDQ} \end{vmatrix} = \begin{vmatrix} \bar{P} & \bar{0} \\ \bar{0} & \bar{1} \end{vmatrix} \begin{vmatrix} \bar{L}_{aa} & \bar{L}_{aR} \\ \bar{L}_{Ra} & \bar{L}_{RR} \end{vmatrix} \begin{vmatrix} \bar{P}^{-1} & \bar{0} \\ \bar{0} & \bar{1} \end{vmatrix} \begin{vmatrix} \bar{P} & \bar{0} \\ \bar{0} & \bar{1} \end{vmatrix} \begin{vmatrix} \bar{i}_{abc} \\ \bar{i}_{FDQ} \end{vmatrix} \tag{4.24}$$

This transformation gives

$$
\begin{vmatrix} \lambda_0 \\ \lambda_d \\ \lambda_q \\ \lambda_F \\ \lambda_D \\ \lambda_Q \end{vmatrix} = \begin{vmatrix} L_0 & 0 & 0 & 0 & 0 & 0 \\ 0 & L_d & 0 & kM_F & kM_D & 0 \\ 0 & 0 & L_q & 0 & 0 & kM_q \\ 0 & kM_F & 0 & L_F & M_R & 0 \\ 0 & kM_D & 0 & M_R & L_D & 0 \\ 0 & 0 & kM_Q & 0 & 0 & L_Q \end{vmatrix} \begin{vmatrix} i_0 \\ i_d \\ i_q \\ i_F \\ i_D \\ i_Q \end{vmatrix}
\tag{4.25}
$$

Define

$$
L_d = L_s + M_s + \frac{3}{2}L_m
\tag{4.26}
$$

$$
L_q = L_s + M_s - \frac{3}{2}L_m
\tag{4.27}
$$

$$
L_0 = L_s - 2M_s
\tag{4.28}
$$

$$
k = \sqrt{\frac{3}{2}}
\tag{4.29}
$$

The inductance matrix is sparse, symmetric, and constant. It decouples the $0dq$ axes.

By transformation to d, q, 0 axes, constant values are obtained and reverse transformation is applied to get to the original stator parameters. This is not discussed in this book. Figure 4.6 shows fundamental concepts of transformation. For further discussions see Reference [3]. The transformation is most important for transient stability models of the generators. Also analogously the induction motors models can also be transformed back and forth.

4.2 INDUCTION MOTORS

4.2.1 Equivalent Circuit

The equivalent circuit of the induction motor, shown in Figure 4.7, is derived in many texts [11]. The power transferred across the air gap is

$$
P_g = I_2^2 \frac{r_2}{s}
\tag{4.30}
$$

Referring to Figure 4.7, r_2 is the rotor resistance, s is the motor slip, and I_2 is the rotor current. Mechanical power developed is the power across the air gap minus copper loss in the rotor, that is,

$$
(1 - s)P_g
\tag{4.31}
$$

Thus, the motor torque T in Newton meters can be written as

$$
T = \frac{1}{\omega_s} I_2^2 \frac{r_2}{s} \approx \frac{1}{\omega_s} \frac{V_1^2(r_2/s)}{\left(R_1 + r_2/s\right)^2 + \left(X_1 + x_2\right)^2}
\tag{4.32}
$$

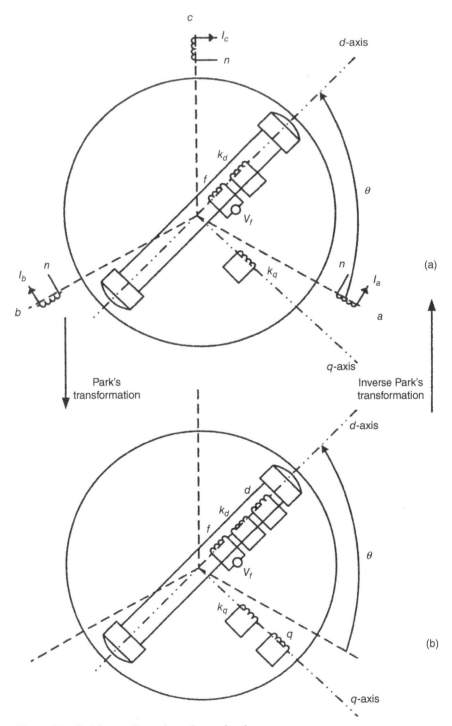

Figure 4.6   Park's transformation—forward and reverse.

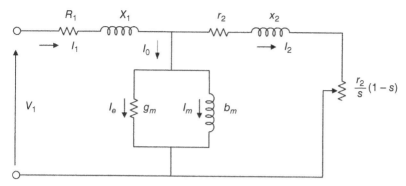

Figure 4.7   Equivalent circuit of an induction motor.

Note that this equation does not contain circuit elements g_m and b_m shown in the equivalent circuit. More accurately, we can write the torque equation as

$$T = \frac{1}{\omega_s} I_2^2 \frac{r_2}{s} \approx \frac{1}{\omega_s} \frac{V_1^2(r_2/s)}{(R_1 + c(r_2/s))^2 + (X_1 + cx_2)^2} \tag{4.33}$$

where c is slightly >1, depending on y_m, r_1 is the stator resistance, V_1 is the terminal voltage, and ω_s is the synchronous angular velocity $= 2\pi f/p$, p being the number of *pairs* of poles. From Equation (4.32), the motor torque varies approximately as the square of the voltage. Also, if the load torque remains constant and the voltage dips, there has to be an increase in the current.

4.2.2 Negative Sequence Impedance

Figure 4.8 shows the negative sequence equivalent circuit of an induction motor. When a negative sequence voltage is applied, the MMF wave in the air gap rotates backward at a slip of 2.0 pu. The slip of the rotor with respect to the backward rotating field is $2 - s$. This results in a retarding torque component and the net motor torque reduces to

$$T = \frac{r_2}{\omega_s} \left(\frac{I_2^2}{s} - \frac{I_{22}^2}{2 - s} \right) \tag{4.34}$$

where I_{22} is the current in the negative sequence circuit.

From equivalent circuits of Figures 4.7 and 4.8, we can write the approximate positive and negative sequence impedances of the motor:

$$Z_1 = \left[\left(R_1 + \frac{r_2}{s} \right)^2 + (X_1 + x_2)^2 \right]^{1/2}$$

$$Z_2 = \left[\left(R_1 + \frac{r_2}{2 - s} \right)^2 + (X_1 + x_2)^2 \right]^{1/2} \tag{4.35}$$

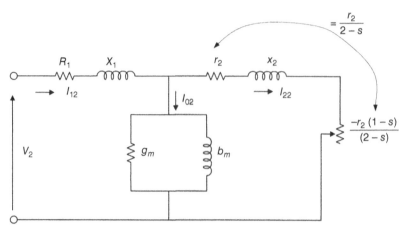

Figure 4.8   Equivalent circuit of an induction motor for negative sequence currents.

Therefore, approximately, the ratio $Z_1/Z_2 = I_s/I_f$, where I_s is the starting current or the locked rotor current of the motor and I_f is the full load current ($s = 1$ at starting). *For an induction motor with a locked rotor current of six times the full-load current, the negative sequence impedance is one-sixth of the positive sequence impedance.* A 5% negative sequence component in the supply system will produce 30% negative sequence current in the motor, which gives rise to additional heating and losses. The rotor resistance will change with respect to high rotor frequency and rotor losses are much higher than the stator losses. A 5% voltage unbalance may give rise to 38% negative sequence current with 50% increase in losses and 40°C higher temperature rise as compared to operation on a balanced voltage with zero negative sequence component. Also, the voltage unbalance is not equivalent to the negative sequence component. The NEMA definition of percentage voltage unbalance is maximum voltage deviation from the average voltage divided by the average voltage as a percentage. Operation above 5% unbalance is not recommended.

The zero sequence impedance of motors, whether the windings are connected in wye or delta formation, is infinite. The motor windings are left ungrounded.

4.2.3 Harmonic Impedances

At fundamental frequency, neglecting magnetizing and loss components, the motor reactance is

$$X_f = X_1 + x_2 \tag{4.36}$$

and the resistance is

$$R_f = R_1 + \frac{r_2}{s} \tag{4.37}$$

This resistance is not the same as used in short-circuit calculations. At harmonic frequencies, the reactance can be directly related to the frequency:

$$X_h = hX_f \tag{4.38}$$

This relation is only approximately correct. The reactance at higher frequencies is reduced due to saturation. The stator resistance can be assumed to vary as the square root of the frequency:

$$R_{1h} = \sqrt{h} \cdot (R_1) \tag{4.39}$$

The harmonic slip is given by

$$s_h = \frac{h-1}{h} \quad \text{for positive sequence harmonics} \tag{4.40}$$

$$s_h = \frac{h+1}{h} \quad \text{for negative sequence harmonics} \tag{4.41}$$

The rotor resistance at harmonic frequencies is

$$r_{2h} = \frac{\sqrt{(1 \pm h)}}{s_h} \tag{4.42}$$

The motor impedance neglecting magnetizing resistance is infinite for triplen harmonics, as the motor windings are not grounded.

The motor impedance at harmonic h is

$$R_1 + jhX_1 + \frac{jhx_m \left(\frac{r_2}{s_h} + jhx_2 \right)}{(r_2/s_h) + jh\left(x_m + x_2 \right)} \tag{4.43}$$

as seen from the input terminals of the motor.

where

$$\begin{aligned} s_h &= 1 - \frac{n}{hn_s} & h &= 3n+1 \\ s_h &= 1 + \frac{n}{hn_s} & h &= 3n-1 \end{aligned} \tag{4.44}$$

Example 4.3 An induction motor is rated 2.3 kV, four-pole, 2500 hp, full-load power factor $= 0.92$, full load efficiency $= 0.93\%$, locked rotor current $=$ six times the full-load current at 20% power factor; calculate its impedance at fundamental frequency and at 5th harmonic.

From the given data, motor full load current $= 547.18$ A

Therefore locked rotor current $= 3283.1$ A at a pf of 20%

$$(R_1 + r_2) + j(X_1 + x_2) = \frac{2.3 \times 10^3}{\sqrt{3}(3283.1)(0.2 - j0.9798)} = 0.081 + j0.396\,\Omega$$

Assume that $\begin{aligned} R_1 &= r_2 = 0.0405\,\Omega \\ X_1 &= x_2 = 0.1980\,\Omega \end{aligned}$

For the induction motor the x_m can be assumed $= 12\,\Omega$ approximately. Also assume full-load slip $= 3\%$.

Then impedance at fundamental frequency is

$$0.0405 + j0.1980 + \frac{j12.0 \left(\frac{0.0405}{0.03} + j0.1980 \right)}{\frac{0.0405}{0.03} + j(12 + 0.1980)}$$

$$= 1.3365 + j0.4340$$

5th harmonic is negative. Then $S_h = 1.20$ (consider $n_s = n$). Again, using the same equation, impedance at 5th harmonic is

$$0.0405 + j4.90 + \frac{j60.0 \left(\frac{0.0405}{1.2} + j4.90 \right)}{\frac{0.0405}{1.2} + j(60 + 4.90)}$$

$$= 0.0695 + j9.43 \ \Omega$$

4.2.4 Zero Sequence Impedance

As per industry practice the neutrals of three-phase motors are not grounded in the United States. Thus, no zero sequence currents can flow.

4.2.5 Terminal Short-Circuit of an Induction Motor

The terminal short-circuit current of an induction motor is given by the equation:

$$i_{ac} + i_{dc} = \frac{E}{X'} e^{-t/T'} + \sqrt{2} \frac{E}{X'} e^{-t/T_{dc}} \tag{4.45}$$

where

$$X' = X_1 + \frac{x_m x_2}{x_m + x_2} \tag{4.46}$$

The short-circuit time constant for decay of ac component is

$$T' = \frac{X'}{\omega r_2} \tag{4.47}$$

The short-circuit time constant for decay of dc component is

$$T_{dc} = \frac{X'}{\omega r_1} \tag{4.48}$$

The total short-circuit current at the instant of fault may rise to 14–16 times the full-load current; but it decays fast. Even for large induction motors the short-current decays to almost 0 in 6–10 cycles.

4.3 STATIC LOADS

Consider a static three-phase load connected in a wye configuration with the neutral grounded through an impedance Z_n. Each phase impedance is Z. The sequence transformation is

$$
\begin{vmatrix} V_a \\ V_b \\ V_c \end{vmatrix} = \begin{vmatrix} Z & 0 & 0 \\ 0 & Z & 0 \\ 0 & 0 & Z \end{vmatrix} \begin{vmatrix} I_a \\ I_b \\ I_c \end{vmatrix} + \begin{vmatrix} I_n Z_n \\ I_n Z_n \\ I_n Z_n \end{vmatrix}
$$

$$
T_s \begin{vmatrix} V_0 \\ V_1 \\ V_2 \end{vmatrix} = \begin{vmatrix} Z & 0 & 0 \\ 0 & Z & 0 \\ 0 & 0 & Z \end{vmatrix} T_s \begin{vmatrix} I_0 \\ I_1 \\ I_2 \end{vmatrix} + \begin{vmatrix} 3 I_0 Z_n \\ 3 I_0 Z_n \\ 3 I_0 Z_n \end{vmatrix} \tag{4.49}
$$

$$
\begin{vmatrix} V_0 \\ V_1 \\ V_2 \end{vmatrix} = T_s^{-1} \begin{vmatrix} Z & 0 & 0 \\ 0 & Z & 0 \\ 0 & 0 & Z \end{vmatrix} T_s \begin{vmatrix} I_0 \\ I_1 \\ I_2 \end{vmatrix} + T_s^{-1} \begin{vmatrix} 3 I_0 Z_n \\ 3 I_0 Z_n \\ 3 I_0 Z_n \end{vmatrix} \tag{4.50}
$$

$$
= \begin{vmatrix} Z & 0 & 0 \\ 0 & Z & 0 \\ 0 & 0 & Z \end{vmatrix} \begin{vmatrix} I_0 \\ I_1 \\ I_2 \end{vmatrix} + \begin{vmatrix} 3 I_0 Z_n \\ 0 \\ 0 \end{vmatrix} = \begin{vmatrix} Z + 3 Z_n & 0 & 0 \\ 0 & Z & 0 \\ 0 & 0 & Z \end{vmatrix} \begin{vmatrix} I_0 \\ I_1 \\ I_2 \end{vmatrix} \tag{4.51}
$$

This shows that the load can be resolved into sequence impedance circuits. This result can also be arrived at by merely observing the symmetrical nature of the circuit.

4.4 HARMONICS AND SEQUENCE COMPONENTS

In a three-phase balanced system under nonsinusoidal conditions, the hth-order harmonic voltage (or current) can be expressed as

$$
V_{ah} = \sum_{h \neq 1} V_h (h \omega_0 t - \theta_h) \tag{4.52}
$$

$$
V_{bh} = \sum_{h \neq 1} V_h (h \omega_0 t - (h \pi / 3) \theta_h) \tag{4.53}
$$

$$
V_{ch} = \sum_{h \neq 1} V_h (h \omega_0 t - (2 h \pi / 3) \theta_h) \tag{4.54}
$$

Based on Equations (4.52)–(4.54), and counterclockwise rotation of the fundamental phasors, we can write

$$
V_a = V_1 \sin \omega t + V_2 \sin 2\omega t + V_3 \sin 3\omega t + V_4 \sin 4\omega t + V_5 \sin 5\omega t + \ldots
$$

$$
V_b = V_1 \sin(\omega t - 120°) + V_2 \sin(2\omega t - 240°) + V_3 \sin(3\omega t - 360°)
$$

$$
+ V_4 \sin(4\omega t - 480°) + V_5 \sin(5\omega t - 600°) + \ldots
$$

$$
= V_1 \sin(\omega t - 120°) + V_2 \sin(2\omega t + 120°) + V_3 \sin 3\omega t + V_4 \sin(4\omega t - 120°)
$$

$$
+ V_5 \sin(5\omega t + 120°) + \ldots
$$

$$V_c = V_1 \sin(\omega t + 120°) + V_2 \sin(2\omega t + 240°) + V_3 \sin(3\omega t + 360°)$$
$$+ V_4 \sin(4\omega t + 480°) + V_5 \sin(5\omega t + 600°) + ...$$
$$= V_1 \sin(\omega t + 120°) + V_2 \sin(2\omega t - 120°) + V_3 \sin 3\omega t + V_4 \sin(4\omega t + 120°)$$
$$+ V_5 \sin(5\omega t - 120°) + ...$$

Under balanced conditions, the hth harmonic (frequency of harmonic = h times the fundamental frequency) of phase b lags h times 120° behind that of the same harmonic in phase a. The hth harmonic of phase c lags h times 240° behind that of the same harmonic in phase a. In the case of triplen harmonics, shifting the phase angles by three times 120° or three times 240° results in co-phasial vectors.

Table 4.4 shows the sequence of harmonics, and the pattern is clearly positive–negative–zero. We can write

$$\text{harmonics of the order } 3h + 1 \text{ have positive sequence,} \tag{4.55}$$

$$\text{harmonics of the order } 3h + 2 \text{ have negative sequence,} \tag{4.56}$$

$$\text{and harmonics of the order } 3h \text{ are of zero sequence.} \tag{4.57}$$

All triplen harmonics generated by nonlinear loads are zero sequence phasors. These add up in the neutral. In a three-phase, four-wire system, with perfectly balanced single-phase loads between the phase and neutral, all positive and negative sequence harmonics will cancel out leaving only the zero sequence harmonics.

In an unbalanced three-phase system, serving single-phase load, the neutral carries zero sequence and the residual unbalance of positive and negative sequence currents. Even harmonics are absent in the line because of phase symmetry and unsymmetrical waveforms will add even harmonics to the phase conductors, for example, half controlled three-phase bridge circuit [12].

REFERENCES

[1] CV Jones. *The Unified Theory of Electrical Machines.* Pergamon Press, 1964.

[2] AT Morgan. *General Theory of Electrical Machines.* London: Heyden & Sons Ltd, 1979.

[3] JC Das. *Transients in Electrical Systems, Analysis Recognition and Mitigation.* New York: McGraw-Hill, 2010.

[4] PM Anderson and A Fouad. *Power System Control and Stability.* New York: IEEE Press, 1991.

[5] NEMA. Large Machines—Synchronous Generators. MG-1, Part 22, 1993.

[6] ANSI. Synchronous Generators, Synchronous Motors and Synchronous Machines in General, 1995. Standard C50.1

[7] ANSI. American Standard Requirements for Cylindrical Rotor Synchronous Generators, 1965. Standard C50.13

[8] MD Ross and JW Batchelor. Operation of non-salient-pole type generators supplying a rectifier load. *AIEE Transactions*, vol. 62, pp. 667–670, 1943.

[9] RH Park. Two reaction theory of synchronous machines, Part I. *AIEE Transactions*, vol. 48, pp. 716–730, 1929.

[10] RH Park. Two reaction theory of synchronous machines, Part II. *AIEE Transactions*, vol. 52, pp. 352–355, 1933.

[11] AE Fitzgerlad, C Kingsley, and A Kusko. *Electrical Machinery*, 4th edition, New York: McGraw-Hill, 1971.

[12] JC Das. *Power System Harmonics and Passive Filter Designs.* New Jersey: IEEE Press, 2015.

FURTHER READING

Adkins, B. *The General Theory of Electrical Machines*. London: Chapman & Hall, 1964.

Anderson, PM. *Analysis of Faulted Power Systems*. Ames, IA: Iowa State University Press, 1973.

Boldea, I. *Synchronous Generators*. Boca Raton, FL: CRC Press, 2005.

Concordia, C. *Synchronous Machines*. New York: John Wiley & Sons, 1951.

Fitzgerald Jr., AE, Umans, SD, and Kingsley, C. *Electrical Machinery*. New York: McGraw-Hill Higher Education, 2002.

Hancock, NN. *Matrix Analysis of Electrical Machinery*. Pergamon Press, 1964.

IEEE Committee Report. Recommended phasor diagram for synchronous machines. *IEEE Transactions on Power Apparatus and Systems*, vol. PAS-88, pp. 1593–1610. 1963.

Park, RH. Two reaction theory of synchronous machines, Part 1. *AIEEE Transactions*, vol. 48, pp. 716–730, 1929.

CHAPTER **5**

THREE-PHASE MODELS OF TRANSFORMERS AND CONDUCTORS

5.1 THREE-PHASE MODELS

The chapter describes three-phase models of two-winding three-phase transformers and conductors. A reader is advised to pursue this chapter along with Chapter 7.

5.2 THREE-PHASE TRANSFORMER MODELS

A three-phase two-winding transformers, total of six windings can be considered a 12-terminal coupled network, consisting of three primary windings and three secondary windings mutually coupled through the transformer core (Figure 5.1). Each winding has some coupling with all the other windings. The short-circuit primitive matrix for this network can be written as

$$
\begin{vmatrix} i_1 \\ i_2 \\ i_3 \\ i_4 \\ i_5 \\ i_6 \end{vmatrix} = \begin{vmatrix} y_{11} & y_{12} & y_{13} & y_{14} & y_{15} & y_{16} \\ y_{21} & y_{22} & y_{23} & y_{24} & y_{25} & y_{26} \\ y_{31} & y_{32} & y_{33} & y_{34} & y_{35} & y_{36} \\ y_{41} & y_{42} & y_{43} & y_{44} & y_{45} & y_{46} \\ y_{51} & y_{52} & y_{53} & y_{54} & y_{55} & y_{56} \\ y_{61} & y_{62} & y_{63} & y_{64} & y_{65} & y_{66} \end{vmatrix} \begin{vmatrix} v_1 \\ v_2 \\ v_3 \\ v_4 \\ v_5 \\ v_6 \end{vmatrix} \tag{5.1}
$$

This ignores tertiary windings. It becomes a formidable problem for calculation if all the Y elements are distinct. Making use of the symmetry the Y matrix can be reduced to

$$
\begin{vmatrix} y_p & -y_m & y_m' & y_m'' & y_m' & y_m'' \\ -y_m & y_s & y_m'' & y_m''' & y_m'' & y_m''' \\ y_m' & y_m'' & y_p & -y_m & y_m' & y_m'' \\ y_m'' & y_m''' & -y_m & y_s & y_m'' & y_m''' \\ y_m' & y_m'' & y_m' & y_m'' & y_p & -y_m \\ y_m'' & y_m''' & y_m'' & y_m''' & -y_m & y_s \end{vmatrix} \tag{5.2}
$$

Understanding Symmetrical Components for Power System Modeling, First Edition. J.C. Das.
© 2017 by The Institute of Electrical and Electronics Engineers, Inc. Published 2017 by John Wiley & Sons, Inc.

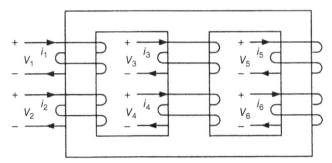

Figure 5.1 Circuit of a grounded wye-delta transformer with voltage and current relations for derivation of connection matrix.

This considers that windings 1, 3, and 5 are primary windings and windings 2, 4, and 6 are secondary windings with appropriate signs for the admittances. The primed elements are all zero if there are no mutual couplings, for example, in the case of three single-phase transformers.

$$
\begin{vmatrix} i_1 \\ i_2 \\ i_3 \\ i_4 \\ i_5 \\ i_6 \end{vmatrix} =
\begin{vmatrix}
y_p & -y_m & 0 & 0 & 0 & 0 \\
-y_m & y_s & 0 & 0 & 0 & 0 \\
0 & 0 & y_p & -y_m & 0 & 0 \\
0 & 0 & -y_m & y_s & 0 & 0 \\
0 & 0 & 0 & 0 & y_p & -y_m \\
0 & 0 & 0 & 0 & -y_m & y_s
\end{vmatrix}
\begin{vmatrix} v_1 \\ v_2 \\ v_3 \\ v_4 \\ v_5 \\ v_6 \end{vmatrix}
$$

Consider a three-phase wye-delta transformer (Figure 5.2). The branch and node voltages in this figure are related by the following connection matrix:

$$
\begin{vmatrix} v_1 \\ v_2 \\ v_3 \\ v_4 \\ v_5 \\ v_6 \end{vmatrix} =
\begin{vmatrix}
1 & 0 & 0 & 0 & 0 & 0 \\
0 & 0 & 0 & 1 & -1 & 0 \\
0 & 1 & 0 & 0 & 0 & 0 \\
0 & 0 & 0 & 0 & 1 & -1 \\
0 & 0 & 1 & 0 & 0 & 0 \\
0 & 0 & 0 & -1 & 0 & 1
\end{vmatrix}
\begin{vmatrix} V_a \\ V_b \\ V_c \\ V_A \\ V_B \\ V_C \end{vmatrix}
\tag{5.3}
$$

or we can write

$$
\bar{v}_{\text{branch}} = \bar{N}\bar{V}_{\text{node}} \tag{5.4}
$$

$\bar{N}$ is the connection matrix.

Also see References [1–4]. Kron's transformation is applied to the connection matrix $\bar{N}$ to obtain the node admittance matrix

$$
\bar{Y}_{\text{node}} = \bar{N}^t \bar{Y}_{\text{prim}} \bar{N} \tag{5.5}
$$

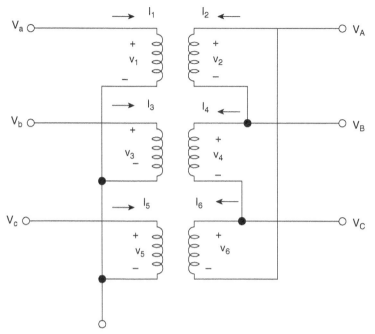

Figure 5.2   Circuit of a grounded wye-delta transformer with voltage and current relations for derivation of connection matrix.

The node admittance matrix is obtained in phase quantities as

$$
\bar{Y}_{\text{node}} =
\begin{vmatrix}
y_s & y'_m & y'_m & -\left(y_m + y''_m\right) & \left(y'_m + y''_m\right) & 0 \\
y'_m & y_s & y'_m & 0 & -\left(y_m + y''_m\right) & \left(y_m + y''_m\right) \\
y'_m & y'_m & y_s & \left(y_m + y''_m\right) & 0 & -\left(y_m + y''_m\right) \\
-\left(y_m + y''_m\right) & 0 & \left(y_m + y''_m\right) & 2\left(y_s - y'''_m\right) & -\left(y_s - y'''_m\right) & -\left(y_s - y'''_m\right) \\
\left(y_m + y''_m\right) & -\left(y_m + y''_m\right) & 0 & -\left(y_s - y'''_m\right) & 2\left(y_s - y'''_m\right) & -\left(y_s - y'''_m\right) \\
0 & \left(y_m + y''_m\right) & -\left(y_m + y''_m\right) & -\left(y_s - y'''_m\right) & -\left(y_s - y'''_m\right) & 2\left(y_s - y'''_m\right)
\end{vmatrix}
$$

(5.6)

The primed y_m vanish when the primitive admittance matrix of three-phase bank is substituted in Equation (5.5). The primitive admittances are considered on per unit basis, and both primary and secondary voltages are 1 per unit. But a wye-delta transformer so obtained must consider a turns ratio of $\sqrt{3}$, so that wye and delta node voltages are still 1.0 per unit. The node admittance matrix can be divided into submatrices as follows:

$$
\bar{Y}_{\text{node}} =
\begin{vmatrix}
\bar{Y}_{\text{I}} & \bar{Y}_{\text{II}} \\
Y_{\text{II}}^t & \bar{Y}_{\text{III}}
\end{vmatrix}
$$

(5.7)

TABLE 5.1 Submatrices of Three-Phase Transformer Connections

Winding Connections		Self-admittance		Mutual Admittance	
Primary	Secondary	Primary	Secondary	Primary	Secondary
Wye–G	Wye–G	$\bar{Y}_{\mathrm{I}}$	$\bar{Y}_{\mathrm{I}}$	$-\bar{Y}_{\mathrm{I}}$	$-\bar{Y}_{\mathrm{I}}$
Wye – G	Wye				
Wye	Wye – G	$\bar{Y}_{\mathrm{II}}$	$\bar{Y}_{\mathrm{II}}$	$-\bar{Y}_{\mathrm{II}}$	$-\bar{Y}_{\mathrm{II}}$
Wye	Wye				
Wye–G	Delta	$\bar{Y}_{\mathrm{I}}$	$\bar{Y}_{11}$	$\bar{Y}_{\mathrm{III}}$	$\bar{Y}_{\mathrm{III}}^{t}$
Wye	Delta	$\bar{Y}_{\mathrm{II}}$	$\bar{Y}_{11}$	$\bar{Y}_{\mathrm{III}}$	$\bar{Y}_{\mathrm{III}}^{t}$
Delta	Wye	$\bar{Y}_{\mathrm{II}}$	$\bar{Y}_{\mathrm{III}}$	$\bar{Y}_{\mathrm{III}}^{t}$	$\bar{Y}_{\mathrm{III}}$
Delta	Wye–G	$\bar{Y}_{\mathrm{II}}$	$\bar{Y}_{11}$	$\bar{Y}_{\mathrm{III}}^{t}$	$\bar{Y}_{\mathrm{III}}$
Delta	Delta	$\bar{Y}_{\mathrm{II}}$	$\bar{Y}_{\mathrm{II}}$	$-\bar{Y}_{\mathrm{II}}$	$-\bar{Y}_{\mathrm{II}}$

Y_{III}^{t} is transpose of Y_{III}.

where each 3×3 submatrix depends on the winding connections, as shown in Table 5.1. The submatrices in this table are defined as follows:

$$\bar{Y}_{\mathrm{I}} = \begin{vmatrix} y_t & 0 & 0 \\ 0 & y_t & 0 \\ 0 & 0 & y_t \end{vmatrix} \quad \bar{Y}_{\mathrm{II}} = \frac{1}{3}\begin{vmatrix} 2y_t & -y_t & -y_t \\ -y_t & 2y_t & -y_t \\ -y_t & -y_t & 2Y_t \end{vmatrix} \quad \bar{Y}_{\mathrm{III}} = \frac{1}{\sqrt{3}}\begin{vmatrix} -y_t & y_t & 0 \\ 0 & -y_t & y_t \\ y_t & 0 & -y_t \end{vmatrix}$$

(5.8)

Here, y_t is the leakage admittance per phase in per unit. Note that all primed y_m are dropped. Normally these primed values are much smaller than the unprimed values. These are considerably smaller in magnitude than the unprimed values and the numerical values of y_s; y_p, and y_m are equal to the leakage impedance y_t obtained by short-circuit test. It is assumed that all three transformer banks are identical.

Table 5.1 can be used as a simplified approach to the modeling of common core-type three phase transformer in an unbalanced system. With more complete information, the model in Equation (5.6) can be used for benefits of accuracy.

5.2.1 Symmetrical Components of Three-Phase Transformers

As stated in earlier chapters, in most cases it is sufficient to assume that the system is balanced. Then, symmetrical component models can be arrived at by using symmetrical component transformations. Continuing with wye-ground-delta transformer of Figure 5.2, first consider the self-admittance matrix. The transformation is

$$\bar{Y}_{012}^{pp} = \bar{T}_s^{-1}\begin{vmatrix} y_p & y_m' & y_m' \\ y_m' & y_p & y_m' \\ y_m' & y_m' & y_p \end{vmatrix}\bar{T}_s = \begin{vmatrix} y_p + 2y_m' & 0 & 0 \\ 0 & y_p - y_m' & 0 \\ 0 & 0 & y_p - y_m' \end{vmatrix}$$

(5.9)

Note that zero sequence admittance is different from the positive and negative sequence impedances. If y_m' is neglected all three are equal.

Similarly,

$$
\bar{Y}^{ss}_{012} = \frac{1}{3}\bar{T}^{-1}_{s}
\begin{vmatrix}
2\left(y_{s}-y'''_{m}\right) & -\left(y_{s}-y'''_{m}\right) & -\left(y_{s}-y'''_{m}\right) \\
-\left(y_{s}-y'''_{m}\right) & 2\left(y_{s}-y'''_{m}\right) & -\left(y_{s}-y'''_{m}\right) \\
-\left(y_{s}-y'''_{m}\right) & -\left(y_{s}-y'''_{m}\right) & 2\left(y_{s}-y'''_{m}\right)
\end{vmatrix}\bar{T}_{s} =
\begin{vmatrix}
0 & 0 & 0 \\
0 & \left(y_{s}-y'''_{m}\right) & 0 \\
0 & 0 & \left(y_{s}-y'''_{m}\right)
\end{vmatrix}
$$

(5.10)

As expected, there is no zero sequence self-admittance in the delta winding.

Finally, the mutual admittance matrix gives

$$
\bar{Y}^{ps}_{012} = \frac{1}{\sqrt{3}}\bar{T}^{-1}_{s}
\begin{vmatrix}
-\left(y_{m}+y''_{m}\right) & 0 & \left(y_{m}+y''_{m}\right) \\
\left(y_{m}+y''_{m}\right) & -\left(y_{m}+y''_{m}\right) & 0 \\
0 & \left(y_{m}+y''_{m}\right) & -\left(y_{m}+y''_{m}\right)
\end{vmatrix}\bar{T}_{s}
$$

$$
=
\begin{vmatrix}
0 & 0 & 0 \\
0 & -\left(y_{m}+y''_{m}\right) < 30^{\circ} & 0 \\
0 & 0 & -\left(y_{m}+y''_{m}\right) < -30^{\circ}
\end{vmatrix}
$$

(5.11)

A phase shift of 30° occurs in the positive sequence network and a negative phase shift of -30° occurs in the negative sequence. No zero sequence currents can occur between wye and delta side of a balanced wye-delta transformer. Based upon above equations the positive, negative, and zero sequence admittances of the transformer are shown in Figure 5.3

If the phase shift is ignored,

- $y'_{m} = y''_{m} = y'''_{m}$, which are zero in a three-phase bank and,
- $y_{p}-y_{m}$ is small admittance, as y_{p} is only slightly $> y_{m}$ and $y_{p}-y_{m}$ and $y_{s}-y_{m}$ as open circuit admittances.

Then the simplified model returns to that shown in Table 5.1.

If the off-nominal tap ratio between primary and secondary windings is $\alpha{:}\beta$, where α and β are the taps on the primary and secondary side, respectively, in per unit, then the submatrices are modified as follows:

- Divide self-admittance of primary matrix by α^{2}
- Divide self-admittance of secondary matrix by β^{2}
- Divide mutual admittance matrices by $\alpha\beta$

Consider a wye-grounded transformer. Then, from Table 5.1,

$$
\bar{Y}^{abc} =
\begin{vmatrix}
\bar{Y}_{I} & -\bar{Y}_{I} \\
-\bar{Y}_{I} & \bar{Y}_{I}
\end{vmatrix}
=
\begin{vmatrix}
y_{t} & 0 & 0 & -y_{t} & 0 & 0 \\
0 & y_{t} & 0 & 0 & -y_{t} & 0 \\
0 & 0 & y_{t} & 0 & 0 & -y_{t} \\
-y_{t} & 0 & 0 & y_{t} & 0 & 0 \\
0 & -y_{t} & 0 & 0 & y_{t} & 0 \\
0 & 0 & -y_{t} & 0 & 0 & y_{t}
\end{vmatrix}
$$

(5.12)

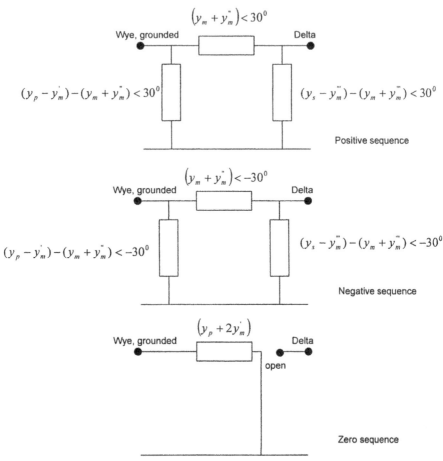

Figure 5.3   Sequence impedances of a wye-grounded delta two-winding transformer.

For off-nominal taps the matrix is modified as

$$
\bar{Y}^{abc} =
\begin{vmatrix}
\dfrac{y_t}{\alpha^2} & 0 & 0 & \dfrac{y_t}{\alpha\beta} & 0 & 0 \\[2mm]
0 & \dfrac{y_t}{\alpha^2} & 0 & 0 & \dfrac{y_t}{\alpha\beta} & 0 \\[2mm]
0 & 0 & \dfrac{y_t}{\alpha^2} & 0 & 0 & \dfrac{y_t}{\alpha\beta} \\[2mm]
\dfrac{y_t}{\alpha\beta} & 0 & 0 & \dfrac{y_t}{\beta^2} & 0 & 0 \\[2mm]
0 & \dfrac{y_t}{\alpha\beta} & 0 & 0 & \dfrac{y_t}{\beta^2} & 0 \\[2mm]
0 & 0 & \dfrac{y_t}{\alpha\beta} & 0 & 0 & \dfrac{y_t}{\beta^2}
\end{vmatrix}
= \left| \bar{Y}_{yg-y} \right| \qquad (5.13)
$$

A three-phase transformer winding connection of delta primary and grounded-wye secondary is commonly used. From Table 5.1, its matrix equation is

$$
\bar{Y}^{abc} =
\begin{vmatrix}
\frac{2}{3}y_t & -\frac{1}{3}y_t & -\frac{1}{3}y_t & -\frac{y_t}{\sqrt{3}} & \frac{y_t}{\sqrt{3}} & 0 \\[2mm]
-\frac{1}{3}y_t & \frac{2}{3}y_t & -\frac{1}{3}y_t & 0 & -\frac{y_t}{\sqrt{3}} & \frac{y_t}{\sqrt{3}} \\[2mm]
-\frac{1}{3}y_t & -\frac{1}{3}y_t & \frac{2}{3}y_t & \frac{y_t}{\sqrt{3}} & 0 & -\frac{y_t}{\sqrt{3}} \\[2mm]
-\frac{y_t}{\sqrt{3}} & 0 & \frac{y_t}{\sqrt{3}} & y_t & 0 & 0 \\[2mm]
\frac{y_t}{\sqrt{3}} & -\frac{y_t}{\sqrt{3}} & 0 & 0 & y_t & 0 \\[2mm]
0 & \frac{y_t}{\sqrt{3}} & -\frac{y_t}{\sqrt{3}} & 0 & 0 & y_t
\end{vmatrix}
\tag{5.14}
$$

where y_t is the leakage reactance of the transformer.

For an off-nominal transformer, the Y matrix is modified as shown below:

$$
\bar{Y}^{abc} =
\begin{vmatrix}
\frac{2}{3}\frac{y_t}{\alpha^2} & \frac{1}{3}\frac{y_t}{\alpha^2} & -\frac{1}{3}\frac{y_t}{\alpha^2} & -\frac{y_t}{\sqrt{3}\alpha\beta} & \frac{y_t}{\sqrt{3}\alpha\beta} & 0 \\[2mm]
-\frac{1}{3}\frac{y_t}{\alpha^2} & \frac{2}{3}\frac{y_t}{\alpha^2} & -\frac{1}{3}\frac{y_t}{\alpha^2} & 0 & -\frac{y_t}{\sqrt{3}\alpha\beta} & \frac{y_t}{\sqrt{3}\alpha\beta} \\[2mm]
-\frac{1}{3}\frac{y_t}{\alpha^2} & -\frac{1}{3}\frac{y_t}{\alpha^2} & \frac{2}{3}\frac{y_t}{\alpha^2} & \frac{y_t}{\sqrt{3}\alpha\beta} & & -\frac{y_t}{\sqrt{3}\alpha\beta} \\[2mm]
-\frac{y_t}{\sqrt{3}\alpha\beta} & 0 & \frac{y_t}{\sqrt{3}\alpha\beta} & \frac{y_t}{\beta^2} & 0 & 0 \\[2mm]
\frac{y_t}{\sqrt{3}\alpha\beta} & -\frac{y_t}{\sqrt{3}\alpha\beta} & 0 & 0 & \frac{y_t}{\beta^2} & 0 \\[2mm]
0 & \frac{y_t}{\sqrt{3}\alpha\beta} & -\frac{y_t}{\sqrt{3}\alpha\beta} & 0 & 0 & \frac{y_t}{\beta^2}
\end{vmatrix}
\tag{5.15}
$$

where α and β are the taps on the primary and secondary sides in per unit.

In a load-flow analysis, the equation of a wye-grounded delta transformer and $a = \beta = 1$ can be written as

$$
\begin{vmatrix} I_A \\ I_B \\ I_C \\ I_a \\ I_b \\ I_c \end{vmatrix} =
\begin{vmatrix}
y_t & 0 & 0 & -\dfrac{1}{\sqrt{3}}y_t & \dfrac{1}{\sqrt{3}}y_t & 0 \\[2mm]
0 & y_t & 0 & 0 & -\dfrac{1}{\sqrt{3}}y_t & \dfrac{1}{\sqrt{3}}y_t \\[2mm]
0 & 0 & y_t & \dfrac{1}{\sqrt{3}}y_t & 0 & -\dfrac{1}{\sqrt{3}}y_t \\[2mm]
-\dfrac{1}{\sqrt{3}}y_t & 0 & \dfrac{1}{\sqrt{3}}y_t & \dfrac{2}{3}y_t & -\dfrac{1}{3}y_t & -\dfrac{1}{3}y_t \\[2mm]
\dfrac{1}{\sqrt{3}}y_t & -\dfrac{1}{\sqrt{3}}y_t & 0 & -\dfrac{1}{3}y_t & \dfrac{2}{3}y_t & -\dfrac{1}{3}y_t \\[2mm]
0 & \dfrac{1}{\sqrt{3}}y_t & -\dfrac{1}{\sqrt{3}}y_t & -\dfrac{1}{3}y_t & -\dfrac{1}{3}y_t & \dfrac{2}{3}y_t
\end{vmatrix}
\begin{vmatrix} V_A \\ V_B \\ V_C \\ V_a \\ V_b \\ V_c \end{vmatrix}
\quad (5.16)
$$

Here, the currents and voltages with capital subscripts relate to primary and those with lower case subscripts relate to secondary. In the condensed form, we will write it as

$$
\bar{I}_{ps} = \bar{Y}_{Y-\Delta}\bar{V}_{ps} \quad (5.17)
$$

Using symmetrical component transformation,

$$
\begin{vmatrix} \bar{I}_p^{012} \\ \bar{I}_s^{012} \end{vmatrix} =
\begin{vmatrix} \bar{T}_s & 0 \\ 0 & \bar{T}_s \end{vmatrix}^{-1}
\bar{Y}_{y-\Delta}
\begin{vmatrix} \bar{T}_s & 0 \\ 0 & \bar{T}_s \end{vmatrix}
\begin{vmatrix} \bar{V}_p^{012} \\ \bar{V}_s^{012} \end{vmatrix}
\quad (5.18)
$$

Expanding,

$$
\begin{vmatrix} \bar{I}_p^{012} \\ \bar{I}_s^{012} \end{vmatrix} =
\begin{vmatrix}
y_t & 0 & 0 & 0 & 0 & 0 \\
0 & y_t & 0 & 0 & y_t < -30° & 0 \\
0 & 0 & y_t & 0 & 0 & y_t < 30° \\
0 & 0 & 0 & 0 & 0 & 0 \\
0 & y_t < 30° & 0 & 0 & y_t & 0 \\
0 & 0 & y_t < -30° & 0 & 0 & y_t
\end{vmatrix}
\begin{vmatrix} \bar{V}_p^{012} \\ \bar{V}_s^{012} \end{vmatrix}
\quad (5.19)
$$

The positive sequence equations are

$$
\begin{aligned}
I_{p1} &= y_t V_{p1} - y_t < -30° V_{s1} \\
I_{s1} &= y_t V_{s1} - y_t < 30° V_{p1}
\end{aligned}
\quad (5.20)
$$

The negative sequence equations are

$$
\begin{aligned}
I_{p2} &= y_t V_{p2} - y_t < 30° V_{s2} \\
I_{s2} &= y_t V_{s2} - y_t < -30° V_2
\end{aligned}
\quad (5.21)
$$

The zero sequence equation is

$$I_{p0} = y_t V_{p0}$$
$$I_{s0} = 0 \tag{5.22}$$

For a balanced system, only the positive sequence component needs to be considered. The power flow on the primary side,

$$
\begin{aligned}
S_{ij} = V_i I_{ij}^* = V_i \left(y_t^* V_i^* - y_t^* < 30^\circ V_j^* \right) \\
= \left[y_t V_i^2 \cos \theta_{yt} - y_t \left| V_i V_j \right| \cos \left(\theta_i - \theta_{yt} - (\theta_j + 30^\circ) \right) \right] \\
+ j \left[-y_t V_i^2 \sin \theta_{yt} - y_t \left| V_i V_j \right| \sin \left(\theta_i - \theta_{yt} - (\theta_j + 30^\circ) \right) \right]
\end{aligned}
\tag{5.23}
$$

and on the secondary side,

$$
\begin{aligned}
S_{ji} = V_j I_{ji}^* = V_j \left(y_t^* V_j^* - y_t^* < -30^\circ V_i^* \right) \\
= \left[y_t V_j^2 \cos \theta_{yt} - y_t \left| V_j V_i \right| \cos \left(\theta_j - \theta_{yt} - (\theta_i - 30^\circ) \right) \right] \\
+ j \left[-y_t V_j^2 \sin \theta_{yt} - y_t \left| V_j V_i \right| \sin \left(\theta_j - \theta_{yt} - (\theta_i - 30^\circ) \right) \right]
\end{aligned}
\tag{5.24}
$$

5.3 CONDUCTORS

A three-phase conductor with mutual coupling between phases and ground wires has an equivalent representation shown in Figures 5.4a and 5.4b, and the following equations are then written for a line segment:

$$
\begin{vmatrix} V_a - V_a' \\ V_b - V_b' \\ V_c - V_c' \end{vmatrix} = \begin{vmatrix} Z_{aa'-g} & Z_{ab'-g} & Z_{ac'-g} \\ Z_{ba'-g} & Z_{bb'-g} & Z_{bc'-g} \\ Z_{ca'-g} & Z_{cb'-g} & Z_{cc'-g} \end{vmatrix} \begin{vmatrix} I_a \\ I_b \\ I_c \end{vmatrix}
\tag{5.25}
$$

In the admittance form, Equation (5.25) can be written as

$$
\begin{vmatrix} I_a \\ I_b \\ I_c \end{vmatrix} = \begin{vmatrix} Y_{aa-g} & Y_{ab-g} & Y_{ac-g} \\ Y_{ba-g} & Y_{bb-g} & Y_{bc-g} \\ Y_{ca-g} & Y_{cb-g} & Y_{cc-g} \end{vmatrix} \begin{vmatrix} V_a - V_a' \\ V_b - V_b' \\ V_c - V_c' \end{vmatrix}
\tag{5.26}
$$

Equation (5.26) can be rearranged as follows:

$$
\begin{aligned}
I_a &= Y_{aa-g}(V_a - V_a') + Y_{ab-g}(V_b - V_b') + Y_{ac-g}(V_c - V_c') \\
&= Y_{aa-g}(V_a - V_a') + Y_{ab-g}(V_a - V_b') + Y_{ac-g}(V_a - V_c') \\
&\quad - Y_{ab-g}(V_a - V_b) + Y_{ac-g}(V_a - V_c) \\
I_b &= Y_{bb-g}(V_b - V_b') + Y_{ba-g}(V_b - V_a') + Y_{bc-g}(V_b - V_c') \\
&\quad - Y_{ba-g}(V_b - V_a) + Y_{bc-g}(V_b - V_c) \\
I_c &= Y_{cc-g}(V_c - V_c') + Y_{cb-g}(V_c - V_b') + Y_{ca-g}(V_c - V_a') \\
&\quad - Y_{cb-g}(V_c - V_b) + Y_{ca-g}(V_c - V_a)
\end{aligned}
\tag{5.27}
$$

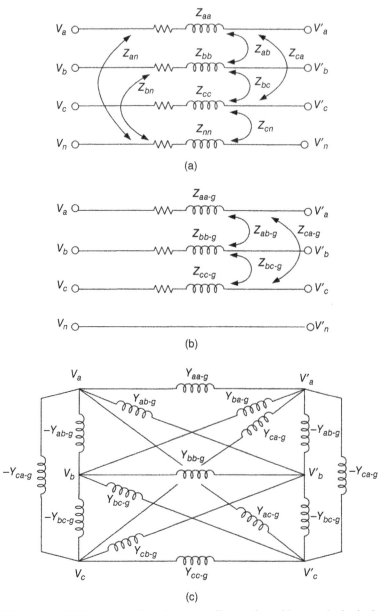

Figure 5.4   (a) Mutual couplings between a line section with ground wire in the impedance form; (b) transformed network in impedance form; (c) equivalent admittance network of a line section.

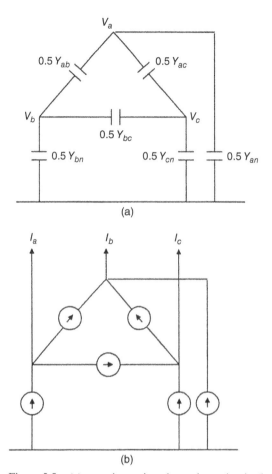

Figure 5.5   (a) capacitance in a three-phase circuit; (b) equivalent current injections.

The same procedure can be applied to nodes V'_a, V'_b, and V'_c. This gives the equivalent series circuit of the line section as shown in Figure 5.4c. The effect of coupling is included in this diagram. Therefore, in the nodal frame we can write the three-phase Π model of a line as

$$
\begin{vmatrix} I_a \\ I_b \\ I_c \\ I'_a \\ I'_b \\ I'_c \end{vmatrix} = \begin{vmatrix} Y^{abc} + \frac{1}{2}Y_{sh} & -Y^{abc} \\ \\ -Y^{abc} & Y^{abc} + \frac{1}{2}Y_{sh} \end{vmatrix} \begin{vmatrix} V_a \\ V_b \\ V_c \\ V'_a \\ V'_b \\ V'_c \end{vmatrix} \tag{5.28}
$$

where

$$
\bar{Y}^{abc} = \bar{Z}^{-1,abc} \tag{5.29}
$$

There is a similarity between the three-phase and single-phase admittance matrix, each element being replaced by a 3×3 matrix.

The shunt capacitance (line charging) can also be represented by current injection. Figure 5.5(a) shows the capacitances of a feeder circuit and Figure 5.5(b) shows current injection. The charging currents are

$$I_a = -\frac{1}{2}[Y_{ab} + Y_{ac} + Y_{an}]V_a + \frac{Y_{ab}}{2}V_b + \frac{Y_{ac}}{2}V_c$$

$$I_b = -\frac{1}{2}[Y_{ab} + Y_{ac} + Y_{an}]V_b + \frac{Y_{ab}}{2}V_a + \frac{Y_{bc}}{2}V_c \qquad (5.30)$$

$$I_c = -\frac{1}{2}[Y_{ab} + Y_{ac} + Y_{an}]V_c + \frac{Y_{ac}}{2}V_a + \frac{Y_{bc}}{2}V_b$$

REFERENCES

[1] TH Chen. Generalized Distribution Analysis System. PhD Dissertation. University of Texas at Arlington, May 1990.
[2] M Chen and WE Dillon. Power system modeling. *Proceedings of the IEEE Power and Systems*, vol. 62, pp. 901–915, 1974.
[3] G Kron. *Tensor Analysis of Networks*. London: McDonald, 1965.
[4] SK Chan. Distribution System Automation. PhD Dissertation. University of Texas at Arlington, 1982.

CHAPTER 6

UNSYMMETRICAL FAULT CALCULATIONS

THE CALCULATIONS of unsymmetrical faults are one of the important applications of symmetrical components. When a fault does not involve all the three phases, we call it an unsymmetrical fault. A three-phase bolted fault means as if the three phases are bolted together with links of zero impedance at the fault point. In Chapter 2 we discussed the nature of sequence networks and how three distinct sequence networks can be constructed as seen from the fault point. Each of these networks can be reduced to a single Thévenin positive, negative, or zero sequence impedance. Only the positive sequence network is active and has a voltage source which is the prefault voltage (Chapter 2). For unsymmetrical fault current calculations, the three separate networks can be connected in a certain manner, depending on the type of fault.

Unsymmetrical fault types involving one or two phases and ground are

- A single line-to-ground fault
- A double line-to-ground fault
- A line-to-line fault

These are called *shunt faults*. A three-phase fault may also involve ground. The unsymmetrical series type faults are

- One-conductor opens
- Two-conductors open

The broken conductors may be grounded on one side or on both sides of the break. An open conductor fault can occur due to operation of a fuse in one of the phases.

Unsymmetrical faults are more common. The most common type is a line-to-ground fault. Approximately 70% of the faults in power systems are single line-to-ground faults.

While applying symmetrical component method to fault analysis, we will ignore the load currents. This makes the positive sequence voltages of all the generators in the system identical and equal to the prefault voltage.

Understanding Symmetrical Components for Power System Modeling, First Edition. J.C. Das.
© 2017 by The Institute of Electrical and Electronics Engineers, Inc. Published 2017 by John Wiley & Sons, Inc.

In the analysis to follow, Z_1, Z_2, and Z_0 are the positive, negative, and zero sequence impedances as seen from the fault point; V_a, V_b, and V_c are the phase-to-ground voltages at the fault point, prior to fault, that is, *if the fault does not exist* and V_1, V_2, and V_0 are corresponding sequence component voltages. Similarly, I_a, I_b, and I_c are the line currents and I_1, I_2, and I_0 their sequence components. A fault impedance of Z_f is assumed in every case. For a bolted fault, $Z_f = 0$.

6.1 LINE-TO-GROUND FAULT

Figure 6.1a shows that phase a of a three-phase system goes to ground through an impedance Z_f. The flow of ground fault current depends on the method of system grounding. A solidly grounded system with zero ground resistance is assumed. There will be some impedance to flow of fault current in the form of impedance of the return ground conductor or the grounding grid resistance. A ground resistance can be added in series with the fault impedance Z_f. The ground fault current must have a return path through the grounded neutrals of generators or transformers. If there is no return path for the ground current, $Z_0 = \infty$ and the ground fault current is zero. This is an obvious conclusion.

Phase a is faulted in Figure 6.1a. As the load current is neglected, currents in phases b and c are zero, and the voltage at the fault point, $V_a = I_a Z_f$. The sequence components of the currents are given by

$$\begin{vmatrix} I_0 \\ I_1 \\ I_2 \end{vmatrix} = \frac{1}{3}\begin{vmatrix} 1 & 1 & 1 \\ 1 & a & a^2 \\ 1 & a^2 & a \end{vmatrix}\begin{vmatrix} I_a \\ 0 \\ 0 \end{vmatrix} = \frac{1}{3}\begin{vmatrix} I_a \\ I_a \\ I_a \end{vmatrix} \tag{6.1}$$

Also,

$$I_0 = I_1 = I_2 = \frac{1}{3}I_a \tag{6.2}$$

$$3I_0 Z_f = V_0 + V_1 + V_2 = -I_0 Z_0 + (V_a - I_1 Z_1) - I_2 Z_2 \tag{6.3}$$

which gives

$$I_0 = \frac{V_a}{Z_0 + Z_1 + Z_2 + 3Z_f} \tag{6.4}$$

The fault current I_a is

$$I_a = 3I_0 = \frac{3V_a}{(Z_1 + Z_2 + Z_0) + 3Z_f} \tag{6.5}$$

This shows that the equivalent fault circuit using sequence impedances can be constructed as shown in Figure 6.1b. In terms of sequence impedances' network blocks, the connections are shown in Figure 6.1c.

This result could also have been arrived at from Figure 6.1b:

$$(V_a - I_1 Z_1) + (-I_2 Z_2) + (-I_0 Z_0) - 3Z_f I_0 = 0$$

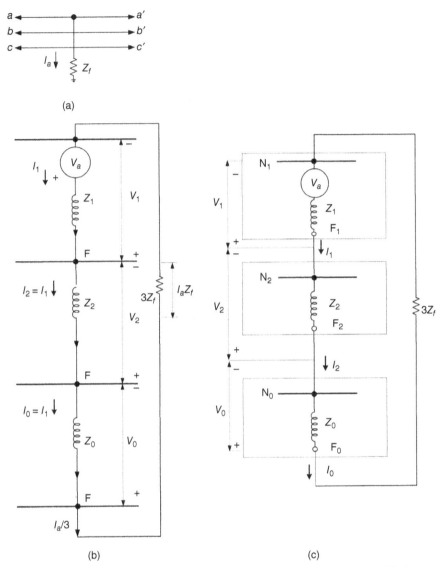

Figure 6.1   (a) Line-to-ground fault in a three-phase system; (b) line-to-ground fault equivalent circuit; (c) sequence network interconnections.

which gives the same Equations (5.4) and (5.5). The voltage of phase b to ground under fault conditions is

$$V_b = a^2 V_1 + a V_2 + V_0$$

$$= V_a \frac{3a^2 Z_f + Z_2(a^2 - a) + Z_0(a^2 - 1)}{(Z_1 + Z_2 + Z_0) + 3Z_f} \qquad (6.6)$$

Similarly, the voltage of phase c can be calculated.

An expression for the ground fault current for use in grounding grid designs and system grounding is as follows:

$$I_a = \frac{3V_a}{(R_0 + R_1 + R_2 + 3R_f + 3R_G) + j(X_0 + X_1 + X_2)} \tag{6.7}$$

where

R_f is the fault resistance
R_G is the resistance of the grounding grid
R_0, R_1, and R_2 are the sequence resistances
X_0, X_1, and X_2 are sequence reactances.

6.2 LINE-TO-LINE FAULT

Figure 6.2a shows a line-to-line fault. A short-circuit occurs between phases b and c, through a fault impedance Z_f. The fault current circulates between phases b and c, flowing back to source through phase b and returning through phase c; $I_a = 0$, $I_b = -I_c$. The sequence components of the currents are

$$\begin{vmatrix} I_0 \\ I_1 \\ I_2 \end{vmatrix} = \frac{1}{3} \begin{vmatrix} 1 & 1 & 1 \\ 1 & a & a^2 \\ 1 & a^2 & a \end{vmatrix} \begin{vmatrix} 0 \\ -I_c \\ I_c \end{vmatrix} = \frac{1}{3} \begin{vmatrix} 0 \\ -a + a^2 \\ -a^2 + a \end{vmatrix} \tag{6.8}$$

From Equation (5.8), $I_0 = 0$ and $I_1 = -I_2$.

$$V_b - V_c = \begin{vmatrix} 0 & 1 & -1 \end{vmatrix} \begin{vmatrix} V_a \\ V_b \\ V_c \end{vmatrix} = \begin{vmatrix} 0 & 1 & -1 \end{vmatrix} \begin{vmatrix} 1 & 1 & 1 \\ 1 & a^2 & a \\ 1 & a & a^2 \end{vmatrix} \begin{vmatrix} V_0 \\ V_1 \\ V_2 \end{vmatrix}$$

$$= \begin{vmatrix} 0 & a^2 - a & a - a^2 \end{vmatrix} \begin{vmatrix} V_0 \\ V_1 \\ V_2 \end{vmatrix} \tag{6.9}$$

Therefore,

$$\begin{aligned} V_b - V_c &= (a^2 - a)(V_1 - V_2) \\ &= (a^2 I_1 + a I_2)Z_f \\ &= (a^2 - a)I_1 Z_f \end{aligned} \tag{6.10}$$

This gives

$$(V_1 - V_2) = I_1 Z_f \tag{6.11}$$

The equivalent circuit is shown in Figures 6.2b and 6.2c.
    Also,

$$I_b = (a^2 - a)I_1 = -j\sqrt{3}I_1 \tag{6.12}$$

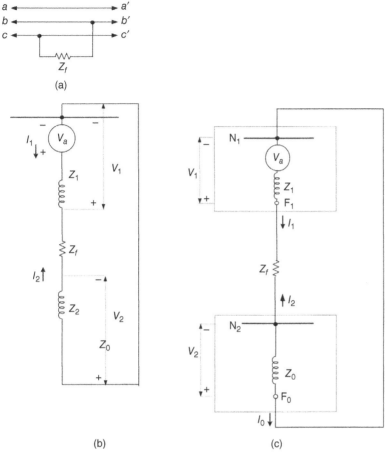

Figure 6.2 (a) Line-to-line fault in a three-phase system; (b) line-to-line fault equivalent circuit; (c) sequence network interconnections.

and

$$I_1 = \frac{V_a}{Z_1 + Z_2 + Z_f} \tag{6.13}$$

The fault current is

$$I_b = -I_c = \frac{-j\sqrt{3}V_a}{Z_1 + Z_2 + Z_f} \tag{6.14}$$

6.3 DOUBLE LINE-TO-GROUND FAULT

A double line-to-ground fault is shown in Figure 6.3a. Phases b and c go to ground through a fault impedance Z_f. The current in the ungrounded phase is zero, that is,

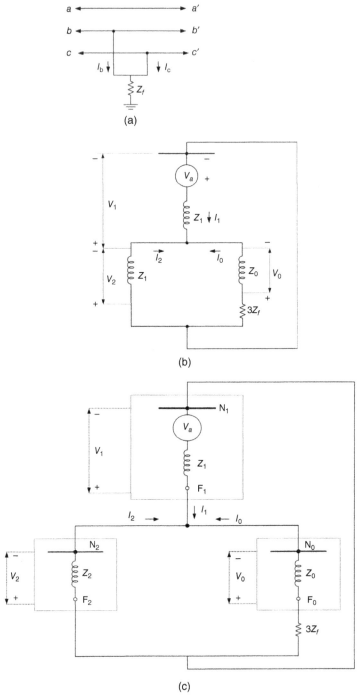

Figure 6.3   (a) Double line-to-ground fault in a three-phase system; (b) double line-to-ground fault equivalent circuit; (c) sequence network interconnections.

$I_a = 0$. Therefore, $I_1 + I_2 + I_0 = 0$.

$$V_b = V_c = (I_b + I_c)Z_f \qquad (6.15)$$

Thus,

$$\begin{vmatrix} V_0 \\ V_1 \\ V_2 \end{vmatrix} = \frac{1}{3}\begin{vmatrix} 1 & 1 & 1 \\ 1 & a & a^2 \\ 1 & a^2 & a \end{vmatrix}\begin{vmatrix} V_a \\ V_b \\ V_b \end{vmatrix} = \frac{1}{3}\begin{vmatrix} V_a + 2V_b \\ V_a + (a + a^2)V_b \\ V_a + (a + a^2)V_b \end{vmatrix} \qquad (6.16)$$

which gives $V_1 = V_2$ and

$$\begin{aligned} V_0 &= \frac{1}{3}(V_a + 2V_b) \\ &= \frac{1}{3}[(V_o + V_1 + V_2) + 2(I_b + I_c)Z_f] \\ &= \frac{1}{3}[(V_0 + 2V_1) + 2(3I_0)Z_f] \\ &= V_1 + 3Z_f I_0 \end{aligned} \qquad (6.17)$$

This gives the equivalent circuit of Figures 6.3b and 6.3c.
    The fault current is

$$\begin{aligned} I_1 &= \frac{V_a}{Z_1 + [Z_2 \parallel (Z_0 + 3Z_f)]} \\ &= \frac{V_a}{Z_1 + \dfrac{Z_2(Z_0 + 3Z_f)}{Z_2 + Z_0 + 3Z_f}} \end{aligned} \qquad (6.18)$$

6.4 THREE-PHASE FAULT

The three phases are short-circuited through equal fault impedances Z_f (Figure 6.4a). The vectorial sum of fault currents is zero, as a symmetrical fault is considered and there is no path to ground.

$$I_0 = I_a + I_b + I_c = 0 \qquad (6.19)$$

As the fault is symmetrical,

$$\begin{vmatrix} V_a \\ V_b \\ V_c \end{vmatrix} = \begin{vmatrix} Z_f & 0 & 0 \\ 0 & Z_f & 0 \\ 0 & 0 & Z_f \end{vmatrix}\begin{vmatrix} I_a \\ I_b \\ I_c \end{vmatrix} \qquad (6.20)$$

The sequence voltages are given by

$$\begin{vmatrix} V_0 \\ V_1 \\ V_2 \end{vmatrix} = [T_s]^{-1}\begin{vmatrix} Z_f & 0 & 0 \\ 0 & Z_f & 0 \\ 0 & 0 & Z_f \end{vmatrix}[T_s]\begin{vmatrix} I_0 \\ I_1 \\ I_2 \end{vmatrix} = \begin{vmatrix} Z_f & 0 & 0 \\ 0 & Z_f & 0 \\ 0 & 0 & Z_f \end{vmatrix}\begin{vmatrix} I_0 \\ I_1 \\ I_2 \end{vmatrix} \qquad (6.21)$$

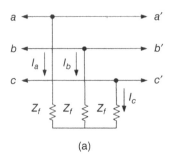

(a)

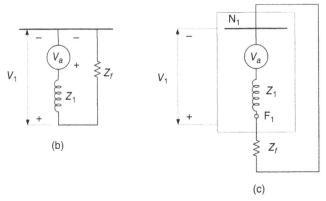

(b)

(c)

Figure 6.4   (a) Three-phase fault; (b) equivalent circuit; (c) sequence network.

This gives the equivalent circuit of Figure 6.4b and 6.4c.

$$I_a = I_1 = \frac{V_a}{Z_1 + Z_f}$$
$$I_b = a^2 I_1$$
$$I_c = a I_1$$

(6.22)

6.5 PHASE SHIFT IN THREE-PHASE TRANSFORMER WINDINGS

6.5.1 Transformer Connections

Transformer windings can be connected in wye, delta, zigzag, or open delta. The transformers may be three-phase units, or three-phase banks can be formed from single-phase units. Autotransformer connections should also be considered. The variety of winding connections is, therefore, large [1]. It is not the intention to describe these connections completely. The characteristics of a connection can be estimated from the vector diagrams of the primary and secondary EMFs. There is a phase shift in the secondary voltages with respect to the primary voltages, depending on the connection. This is of importance when paralleling transformers. A vector diagram of the transformer connections can be constructed based on the following:

1. The voltages of primary and secondary windings on the same leg of the transformer are in opposition, while the induced EMFs are in the same direction.

2. The induced EMFs in three phases are equal, balanced, and displaced mutually by a one-third period in time. These have a definite phase sequence.

Delta-wye connections are discussed, as these are most commonly used. Figure 6.5 shows polarity markings and connections of delta-wye transformers. For all liquid immersed transformers, the polarity is subtractive according to ANSI (American National Standard Institute) standard [2]. Two-winding transformers have their windings designated as high voltage (H) and low voltage (X). Transformers with more than two windings have their windings designated as H, X, Y, and Z. External

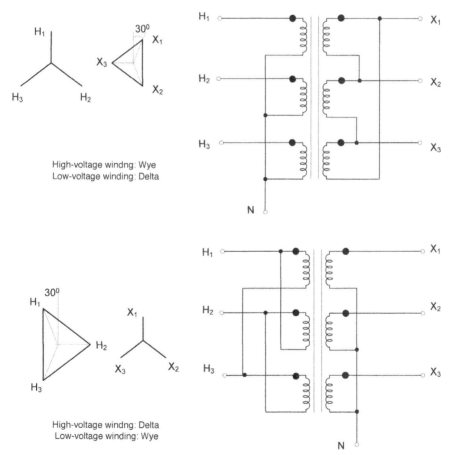

Figure 6.5   Winding connections and phase displacements of voltage vectors for transformers; (a) high-voltage winding in wye and low-voltage winding in delta; (b) high-voltage winding in delta and low-voltage winding in wye connection.

terminals are distinguished from each other by marking with a capital letter, followed by a subscript number, that is, H_1, H_2, and H_3.

6.5.2 Phase Shifts in Winding as per Standards

The angular displacement of a polyphase transformer is the time angle expressed in degrees between the line-to-neutral voltage of the reference identified terminal and the line-to-neutral voltage of the corresponding identified low-voltage terminal. In Figure 6.5a, wye-connected side voltage vectors lead the delta-connected side voltage vectors by 30°, for counterclockwise rotation of phasors. In Figure 6.5b, the delta-connected side leads the wye-connected side by 30°. For transformers manufactured according to the ANSI/IEEE (Institute of Electrical and Electronics Engineers, Inc., USA), standard [3], the *low-voltage side, whether in wye or delta* connection, has a phase shift of 30° lagging with respect to the high-voltage side phase-to-neutral voltage vectors. Figure 6.6 shows ANSI/IEEE [3] transformer connections and a phasor diagram of the delta side and wye side voltages. These relations and phase displacements are applicable to positive sequence voltages.

The International Electrotechnical Commission (IEC) allocates vector groups, giving the type of phase connection and the *angle of advance* turned though in passing from the vector representing the high-voltage side EMF to that representing the low-voltage side EMF at the corresponding terminals. The angle is indicated much like the hands of a clock, the high-voltage vector being at 12 o'clock (zero) and the corresponding low-voltage vector being represented by the hour hand. The total rotation corresponding to hour hand of the clock is 360°. Thus, Dy11 and Yd11 symbols specify 30° lead (11 being the hour hand of the clock) and Dy1 and Yd1 signify 30° lag. Figure 6.7 shows some IEC vector groups of transformers and their winding connections.

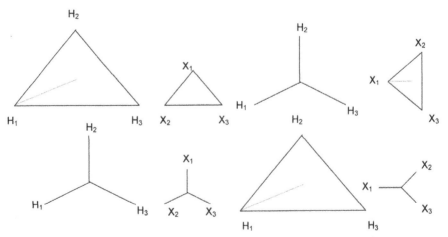

Figure 6.6   Phase displacements and terminal markings in three-phase transformers according to ANSI/IEEE standard.

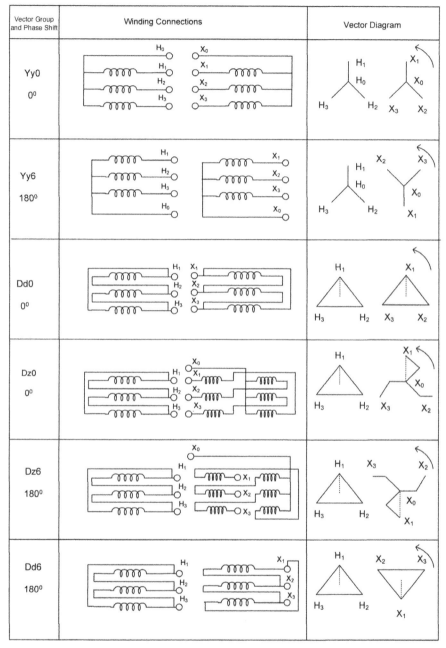

Figure 6.7   Transformer vector groups, winding connections, and vector (phasor) diagrams.

Vector Group and Phase Shift	Winding Connections	Vector Diagrams
Dy1 -30^0	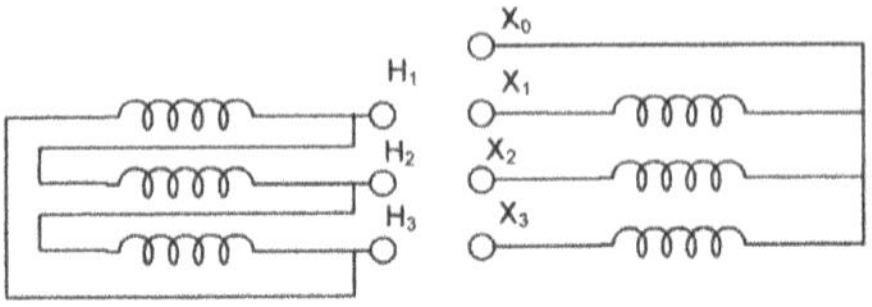	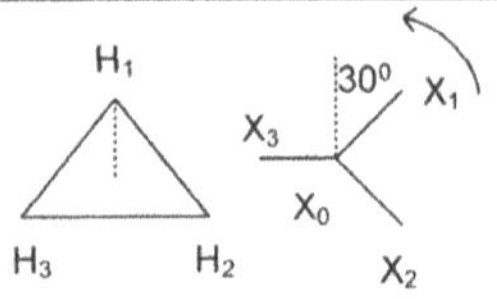
Yd1 -30^0	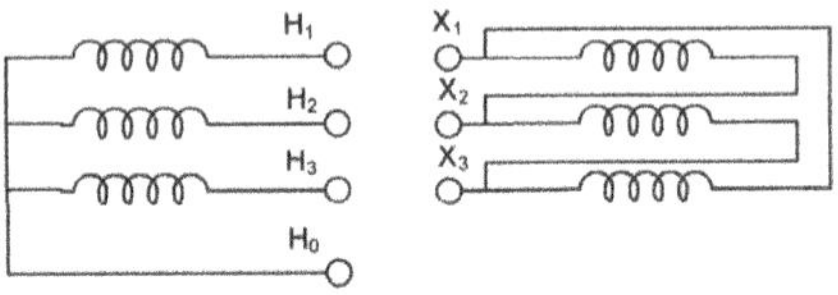	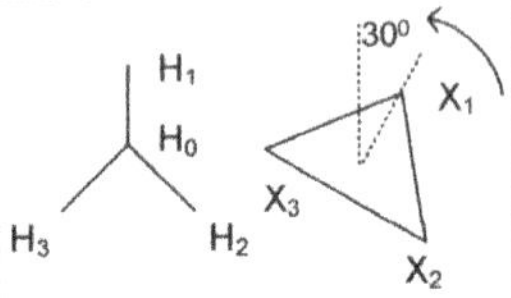
Dy11 $+30^0$	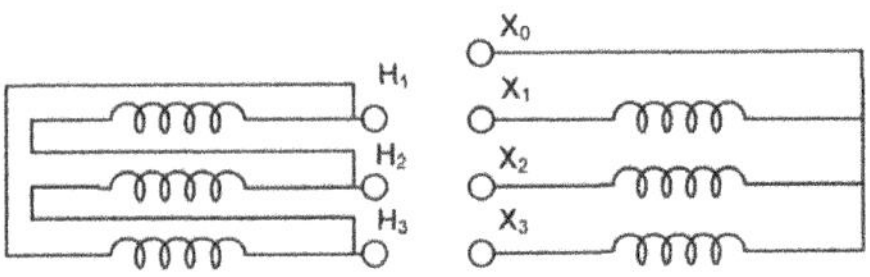	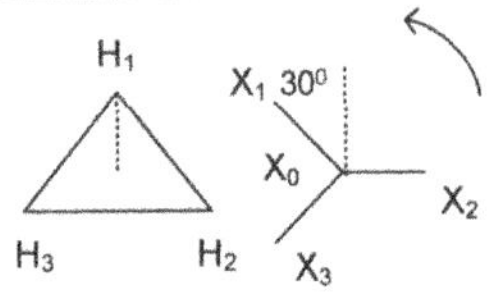
Yd11 $+30^0$	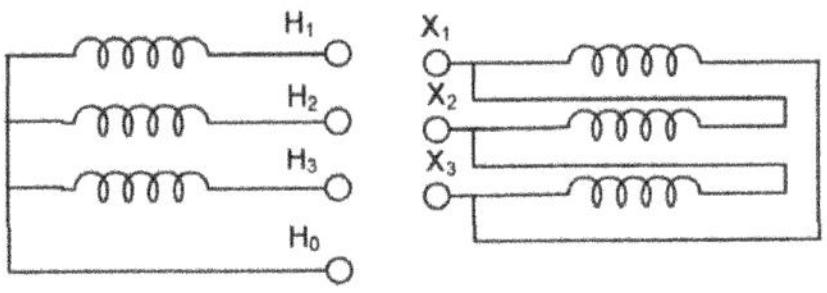	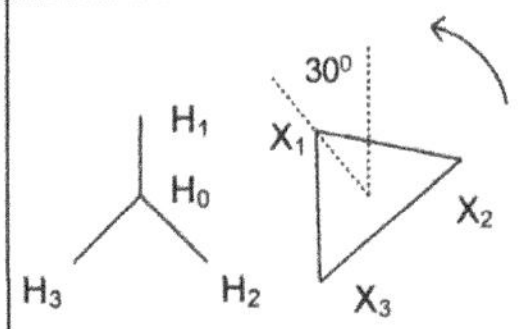
Yz1 -30^0	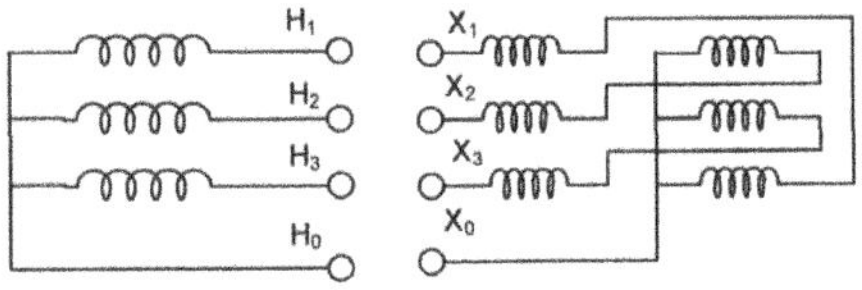	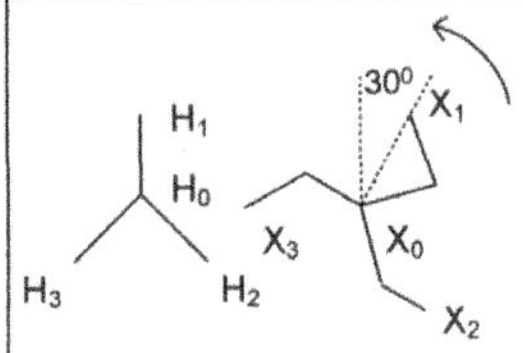
Yz11 $+30^0$	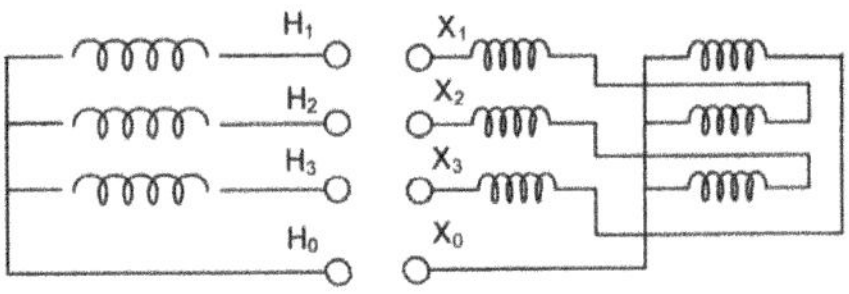	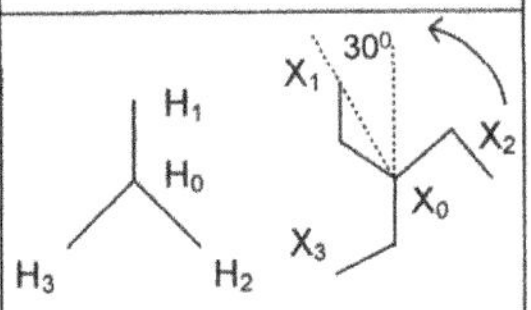

Figure 6.7   (*Continued*)

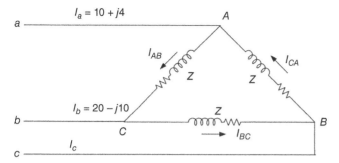

Figure 6.8   Balanced delta connected load on an unbalanced three-phase power supply.

6.5.3 Phase Shift for Negative Sequence Components

The phase shifts described earlier are applicable to positive sequence voltages or currents. If a voltage of negative phase sequence is applied to a delta-wye connected transformer, the phase angle displacement will be equal to the positive sequence phasors, but in the opposite direction. Therefore, when the positive sequence currents and voltages on one side lead the positive sequence current and voltages on the other side by 30°, the corresponding negative sequence currents and voltages will lag by 30°. In general, if the positive sequence voltages and currents on one side lag the positive sequence voltages and currents on the other side by 30°, the negative sequence voltages and currents will lead by 30°.

Example 6.1   Consider a balanced three-phase delta load connected across an unbalanced three-phase supply system, as shown in Figure 6.8. The currents in lines a and b are given.

   The currents in the delta-connected load and also the symmetrical components of line and delta currents are required to be calculated. From these calculations, the phase shifts of positive and negative sequence components in delta windings and line currents can be established.

   The line current in c is given by

$$I_c = -(I_a + I_b)$$
$$= -30 + j6.0A$$

The currents in delta windings are

$$I_{AB} = \frac{1}{3}(I_a - I_b) = -3.33 + j4.67 = 5.735 < 144.51°A$$

$$I_{BC} = \frac{1}{3}(I_b - I_c) = 16.67 - j5.33 = 17.50 < -17.7°A$$

$$I_{CA} = \frac{1}{3}(I_c - I_a) = -13.33 + j0.67 = 13.34 < 177.12°A$$

Calculate the sequence component of the currents I_{AB}. This calculation gives

$$I_{AB1} = 9.43 < 89.57° \text{A}$$
$$I_{AB2} = 7.181 < 241.76° \text{A}$$
$$I_{AB0} = 0 \text{A}$$

Calculate the sequence component of current I_a. This calculation gives

$$I_{a1} = 16.33 < 59.57° \text{A}$$
$$I_{a2} = 12.437 < 271.76° \text{A}$$
$$I_{a0} = 0 \text{A}$$

This shows that the positive sequence current in the delta winding is $1/\sqrt{3}$ times the line positive sequence current, and the phase displacement is +30°, that is,

$$I_{AB1} = 9.43 < 89.57° = \frac{I_{a1}}{\sqrt{3}} < 30° = \frac{16.33}{\sqrt{3}} < (59.57° + 30°) \text{A}$$

The negative sequence current in the delta winding is $1/\sqrt{3}$ times the line negative sequence current, and the phase displacement is −30°, that is,

$$I_{AB2} = 7.181 < 241.76° = \frac{I_{a2}}{\sqrt{3}} < -30° = \frac{12.437}{\sqrt{3}} < (271.76° - 30°) \text{A}$$

This example illustrates that the negative sequence currents and voltages undergo a phase shift which is the reverse of the positive sequence currents and voltages.

The relative magnitudes of fault currents in two winding transformers for secondary faults are shown in Figure 6.9, on a per unit basis. The reader can verify the fault current flows shown in this figure.

6.6 UNSYMMETRICAL LONG HAND FAULT CALCULATIONS

We will demonstrate the procedure of calculations of unsymmetrical fault in a simple distribution system configuration with an example.

Example 6.2   The calculations using symmetrical components can best be illustrated with an example. Consider a subtransmission system as shown in Figure 6.10. A 13.8-kV generator G_1 voltage is stepped up to 138 kV. At the consumer end, the voltage is stepped down to 13.8 kV, and generator G_2 operates in synchronism with the supply system. Bus B has a 10,000-hp motor load. A line-to-ground fault occurs at bus B. It is required to calculate the fault current distribution throughout the system and also the fault voltages. The resistance of the system components is ignored in the calculations. This is rather a simple system. Note that practically the power system studies may run into more than 1000 buses; even a medium size industrial distribution system with a load demand of 70 MW, may have more than 600 buses to modal.

Fault type	Primary	Secondary

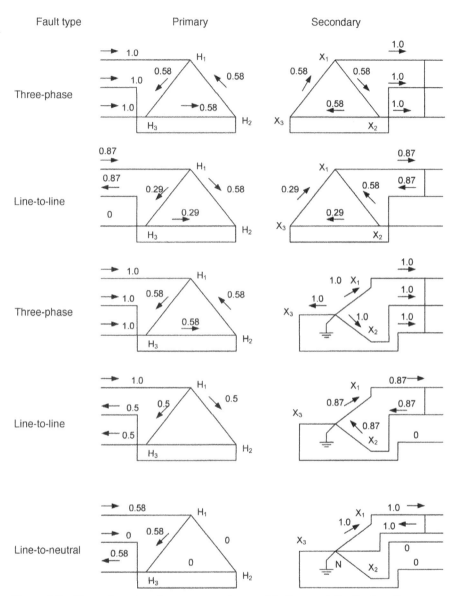

Figure 6.9   Three-phase transformer connections and fault current distributions for secondary faults.

Impedance Data

The impedance data for the system components are shown in Table 6.1. Generators G_1 and G_2 are shown solidly grounded, which will not be the case in a practical installation. A high-impedance grounding system is used by utilities for grounding generators in step-up transformer configurations. Generators in industrial facilities, directly connected to the load buses are low-resistance grounded, and the ground

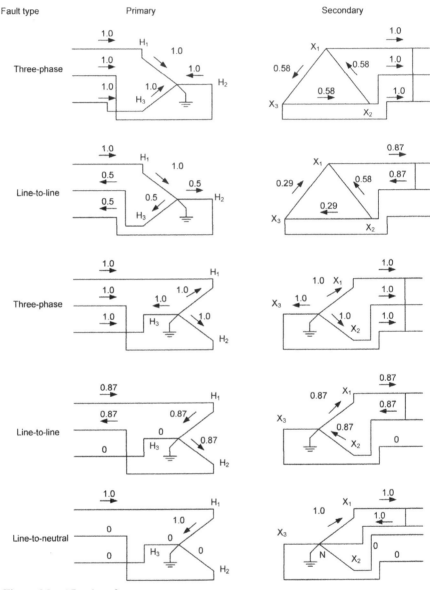

Figure 6.9 *(Continued)*

fault currents are limited to 200–400 A. The simplifying assumptions in the example are not applicable to a practical installation, but clearly illustrate the procedure of calculations.

The first step is to examine the given impedance data. Generator-saturated sub-transient reactance is used in the short-circuit calculations and this is termed positive sequence reactance; 138-kV transmission line reactance is calculated from the

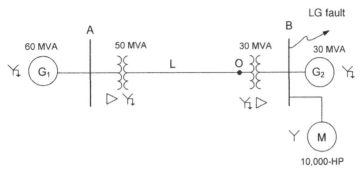

Figure 6.10 A single line diagram of a small power system for Example 6.2.

TABLE 6.1 Impedance Data for Example 6.2

Equipment	Description	Impedance Data	Per Unit Impedance 100-MVA Base
G_1	13.8-kV, 60-MVA, 0.85 power factor generator	Subtransient reactance = 15% Transient reactance = 20% Zero sequence reactance = 8% Negative sequence reactance = 16.8%	$X_1 = 0.25$ $X_2 = 0.28$ $X_0 = 0.133$
T_1	13.8–138 kV step-up transformer, 50/84 MVA, delta-wye connected	$Z = 9\%$ on 50-MVA base	$X_1 = X_2 = X_0$ $= 0.18$
L_1	Transmission line, 5 miles long, 266.8 kcmil, ACSR	Conductors at 15 ft (4.57 m) equivalent spacing	$X_1 = X_2 = X_0$ $= 0.15$
T_2	138–13.2 kV, 30-MVA step-down transformer, wye-delta connected	$Z = 8\%$	$X_1 = X_2 = X_0$ $= 0.24$
G_2	13.8-kV, 30-MVA, 0.85 power factor generator	Subtransient reactance = 11% Transient reactance = 15% Zero sequence reactance = 6% Negative sequence reactance = 16.5%	$X_1 = 0.37$ $X_2 = 0.55$ $X_0 = 0.20$
M	10,000-hp induction motor load	Locked rotor reactance = 16.7% on motor base kVA (consider hp ≈ 1 kVA)	$X_1 = 1.67$ $X_2 = 1.80$ $X_0 = \infty$

Resistances are neglected in the calculations.
kcmil: Kilo-circular mils, same as MCM.
ACSR: Aluminum conductor steel reinforced.

given data for conductor size and equivalent conductor spacing. The zero sequence impedance of the transmission line cannot be completely calculated from the given data and is estimated on the basis of certain assumptions, that is, a soil resistivity of 100 Ω m.

Compiling the impedance data for the system under study from the given parameters, from manufacturers' data, or by calculation and estimation can be time consuming. Most computer-based analysis programs have extensive data libraries and companion programs for calculation of system impedance data and line constants, which has partially removed the onus of generating the data from step-by-step analytical calculations. Needless to emphasis that the accuracy of the results of calculation depends upon the accuracy of the input data.

Next, the impedance data are converted to a common MVA base. A familiarity with the per unit system is assumed. The voltage transformation ratio of transformer T_2 is 138–13.2 kV, while a bus voltage of 13.8 kV is specified, which should be considered in transforming impedance data on a common MVA base. Table 6.1 shows raw impedance data and their conversion into sequence impedances.

Long Hand Calculations

For a single line-to-ground fault at bus B, the sequence impedance network connections are shown in Figure 6.11, with the impedance data for components clearly marked. This figure is based on the fault equivalent circuit shown in Figure 6.1b, with fault impedance $Z_f = 0$. The calculation is carried out per unit, and the units are not stated in every step of the calculation.

The positive sequence impedance to the fault point is

$$Z_1 = \frac{j(0.25 + 0.18 + 0.04 + 0.24) \times \dfrac{j0.37 \times j1.67}{j(0.37 + 1.67)}}{j(0.25 + 0.18 + 0.04 + 0.24) + \dfrac{j0.37 \times j1.67}{j(0.37 + 1.67)}}$$

This gives $Z_1 = j0.212$.

$$Z_2 = \frac{j(0.28 + 0.18 + 0.04 + 0.24) \times \dfrac{j0.55 \times j1.8}{j(0.55 + 1.8)}}{j(0.28 + 0.18 + 0.04 + 0.24) + \dfrac{j0.55 \times j1.8}{j(0.55 + 1.8)}}$$

This gives $Z_2 = j0.266$.

$Z_0 = j0.2$. Therefore,

$$I_1 = \frac{E}{Z_1 + Z_2 + Z_0} = \frac{1}{j0.212 + j0.266 + j0.2} = -j1.475$$
$$I_2 = I_0 = -j1.475$$
$$I_a = I_0 + I_1 + I_2 = 5(-j1.475) = -j4.425 \text{ pu}$$

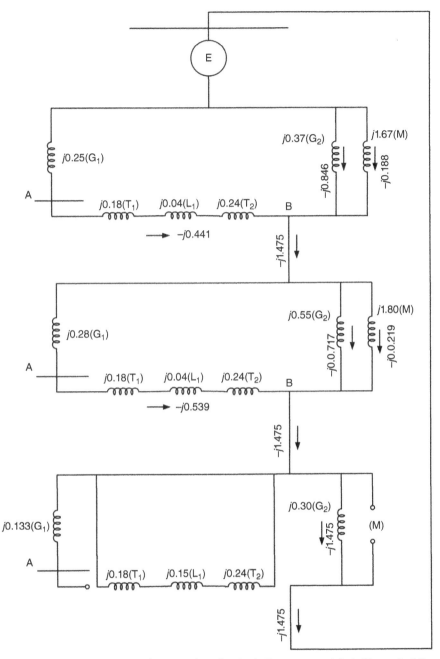

Figure 6.11   Sequence network connections for single line-to-ground fault (Example 6.2).

In terms of actual values this is equivalent to 1.851 kA. The fault currents in phases b and c are zero:

$$I_b = I_c = 0$$

The sequence voltages at a fault point can now be calculated:

$$V_0 = -I_0 Z_0 = j1.475 \times j0.2 = -0.295$$
$$V_2 = -I_2 Z_2 = j1.475 \times j0.266 = -0.392$$
$$V_1 = E - I_1 Z_1 = I_1(Z_0 + Z_2) = 1 - (-j1.475 \times j0.212) = 0.687$$

A check of the calculation can be made at this stage; the voltage of the faulted phase at fault point $B = 0$:

$$V_a = V_0 + V_1 + V_2 = -0.295 - 0.392 + 0.687 = 0$$

The voltages of phases b and c at the fault point are

$$
\begin{aligned}
V_b &= V_0 + aV_1 + a^2 V_2 \\
&= V_0 - 0.5(V_1 + V_2) - j0.866(V_1 - V_2) \\
&= -0.295 - 0.5(0.687 - 0.392) - j0.866(0.687 + 0.392) \\
&= -0.4425 - j0.9344
\end{aligned}
$$
$$|V_b| = 1.034 \,\text{pu}$$

Similarly,

$$V_c = V_0 - 0.5(V_1 + V_2) + j0.866(V_1 - V_2)$$
$$= -0.4425 + j0.9344$$
$$|V_c| = 1.034 \,\text{pu}$$

The distribution of the sequence currents in the network is calculated from the known sequence impedances. The positive sequence current contributed from the right side of the fault, that is., by G_2 and motor M is

$$-j1.475 \frac{j(0.25 + 0.18 + 0.04 + 0.24)}{j(0.25 + 0.18 + 0.04 + 0.24) + \dfrac{j0.37 \times j1.67}{j(0.37 + 1.67)}}$$

This gives $-j1.0338$. This current is composed of two components, one from the generator G_2 and the other from the motor M. The generator component is

$$(-j1.0338)\frac{j1.67}{j(0.37 + 1.67)} = -j0.8463$$

The motor component is similarly calculated and is equal to $-j0.1875$.

The positive sequence current from the left side of bus B is

$$-j1.475 \frac{\dfrac{j0.37 \times j1.67}{j(0.37 + 1.67)}}{j(0.25 + 0.18 + 0.04 + 0.24) + \dfrac{j0.37 \times j1.67}{j(0.37 + 1.67)}}$$

This gives $-j0.441$. The currents from the right side and the left side should sum to $-j1.475$. This checks the calculation accuracy.

The negative sequence currents are calculated likewise and are as follows:

In generator $G_2 = -j0.7172$

In motor $M = -j0.2191$

From left side, Bus B $= -j0.5387$

From right side $= -j0.9363$

The results are shown in Figure 6.11. Again, verify that the vector summation at the junctions confirms the accuracy of calculations.

Currents in Generator G_2

$$I_a(G_2) = I_1(G_2) + I_2(G_2) + I_0(G_2)$$
$$= -j0.8463 - j0.7172 - j1.475$$
$$= -j3.0385$$
$$|I_a(G_2)| = 3.0385 \text{ pu}$$
$$I_b(G_2) = I_0 - 0.5(I_1 + I_2) - j0.866(I_1 - I_2)$$
$$= -j1.475 - 0.5(-j0.8463 - j0.7172) - j0.866(-j0.8463 + j0.7172)$$
$$= -0.1118 - j0.6933$$
$$|I_b(G_2)| = 0.7023 \text{ pu}$$
$$I_c(G_2) = I_0 - 0.5(I_1 + I_2) + j0.866(I_1 - I_2)$$
$$= 0.1118 - j0.6933$$
$$|I_c(G_2)| = 0.7023 \text{ pu}$$

This large unbalance is noteworthy. It gives rise to increased thermal effects due to negative sequence currents and results in overheating of the generator rotor. A generator will be tripped quickly on negative sequence currents.

Currents in Motor M. The zero sequence current in the motor is zero, as the motor wye connected windings are not grounded as per industrial practice in the United States. Thus,

$$I_a(M) = I_1(M) + I_2(M)$$
$$= -j0.1875 - j0.2191$$
$$= -j0.4066$$
$$|I_a(M)| = 0.4066 \text{ pu}$$

$$I_b(M) = -0.5(-j0.4066) - j0.866(0.0316) = 0.0274 + j0.2033$$
$$I_c(M) = -0.0274 + j0.2033$$
$$|I_b(M)| = |I_c(M)| = 0.2051 \text{ pu}$$

The summation of the line currents in the motor M and generator G_2 are

$$I_a(G_2) + I_a(M) = -j3.0385 - j0.4066 = -j3.4451$$
$$I_b(G_2) + I_b(M) = -0.118 - j0.6993 + 0.0274 + j0.2033 = -0.084 - j0.490$$
$$I_c(G_2) + I_c(M) = 0.1118 - j0.6933 - 0.0274 + j0.2033 = 0.084 - j0.490$$

Currents from the left side of the bus B are

$$I_a = -j0.441 - j0.5387$$
$$= -j0.98$$
$$I_b = -0.5(-0.441 - j0.5387) - j0.866(-0.441 + j0.5387)$$
$$= 0.084 + j0.490$$
$$I_c = -0.084 + j0.490$$

These results are consistent as the sum of currents in phases b and c at the fault point from the right and left side is zero and the summation of phase a currents gives the total ground fault current at $b = -j4.425$. The distribution of currents is shown in a three-line diagram (Figure 6.12).

Continuing with the example, the currents and voltages in the transformer T_2 windings are calculated. We should correctly apply the phase shifts for positive and negative sequence components when passing from delta secondary to wye primary of the transformer. The positive and negative sequence current on the wye side of transformer T_2 are

$$I_{1(p)} = I_1 < 30° = -j0.441 < 30° = 0.2205 - j0.382$$
$$I_{2(p)} = I_2 < -30° = -j0.5387 < -30° = -0.2695 - j0.4668$$

Also, the zero sequence current is zero. The primary currents are

$$I_{a(p)} = I_{1(p)} + I_{2(p)}$$
$$= 0.441 < 30° + 0.5387 < -30° = -0.049 - j0.8487$$
$$I_{b(p)} = a^2 I_{1(p)} + a I_{2(p)} = -0.0979$$
$$I_{c(p)} = a I_{1(p)} + a^2 I_{2(p)} = -0.049 - j0.8487$$

Currents in the lines on the delta side of the transformer T_1 are similarly calculated. The positive sequence component, which underwent a 30° positive shift from delta to wye in transformer T_2, undergoes a −30° phase shift; as for an ANSI connected transformer it is the low-voltage vectors which lag the high-voltage side vectors. Similarly, the negative sequence component undergoes a positive phase shift. The currents on the delta side of transformers T_1 and T_2 are identical in amplitude and phase. Note that 138 kV line is considered lossless. Figure 6.12 shows the distribution of currents throughout the distribution system.

The voltage on the primary side of transformer T_2 can be calculated. The voltages undergo the same phase shifts as the currents. Positive sequence voltage is the

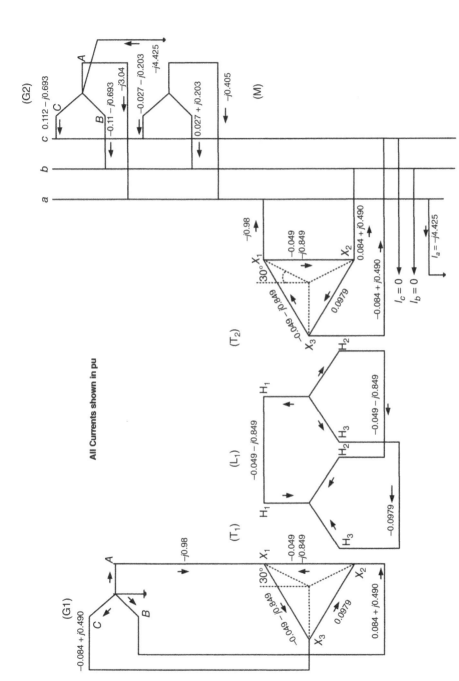

Figure 6.12 Fault current distribution shown in a three-line diagram (Example 6.2).

base fault positive sequence voltage, phase shifted by 30° (positive) minus the voltage drop in transformer reactance due to the positive sequence current:

$$V_1(p) = 1.0 < 30° - jI_{1(p)}X_2$$
$$= 1.0 < 30° - (j0.441 < 30°)(j0.24)$$
$$= 0.9577 + j0.553$$
$$V_2(p) = 0 - I_{2(p)}X_2$$
$$= -(0.539 < -30°)(j0.24)$$
$$= 0.112 - j0.0647$$

Thus,

$$V_{a(p)} = 0.9577 + j0.553 + 0.112 - j0.0647 = 1.0697 + j0.4883 = 1.17 < 24.5°$$
$$V_{b(p)} = -0.5(V_{1(p)} + V_{2(p)}) - j0.866(V_{1(p)} - V_{2(p)})$$
$$= -j0.9763$$
$$V_{c(p)} = 0.5(V_{1(p)} + V_{2(p)}) - j0.866(V_{2(p)} - V_{1(p)})$$
$$= 1.0697 + j0.4883 = 1.17 < 155.5°$$

Note the voltage unbalance caused by the fault.

Concept 6.1 *The "long way" of calculation using symmetrical components, illustrated by the example, shows that, even for simple systems, the calculations are tedious and lengthy. For large networks consisting of thousands of branches and nodes these are impractical. There is an advantage in the hand calculations, in the sense that verification is possible at each step and the results can be correlated with the expected final results. For large systems, matrix methods and digital simulation of the systems are invariable.*

The matrix method is further discussed in this chapter in a following section 6.8. The sparse matrix techniques and matrix manipulations are not discussed. An interested reader may see References [4–8].

6.7 OPEN CONDUCTOR FAULTS

Symmetrical components can also be applied to the study of open conductor faults. These faults are in series with the line and are called series faults. One or two conductors may be opened due to mechanical damage or by operation of fuses on unsymmetrical faults.

6.7.1 Two Conductor Open Fault

Consider that conductors of phases b and c are open-circuited. The currents in these conductors then go to zero.

$$I_b = I_c = 0 \tag{6.23}$$

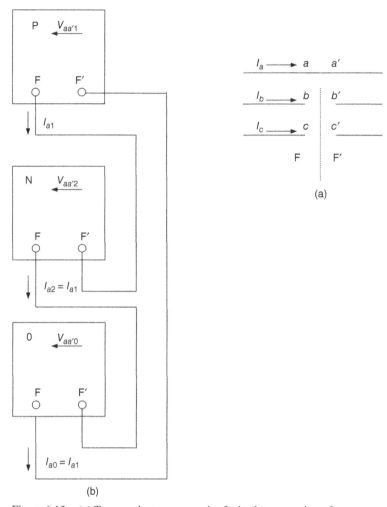

Figure 6.13   (a) Two conductor open series fault; (b) connection of sequence networks.

The voltage across the unbroken phase conductor is zero, at the point of break (Figure 6.13a).

$$V_{a0} = V_{ao1} + V_{a02} + V_{a0} = 0$$
$$I_{a1} = I_{a2} = I_{a0} = \frac{1}{3}I_a$$

(6.24)

This suggests that sequence networks can be connected in series as shown in Figure 6.13b.

6.7.2 One Conductor Open Fault

Now consider that phase a conductor is broken (Figure 6.14a)

$$I_a = 0 \quad V_{b0} - V_{c0} = 0$$

(6.25)

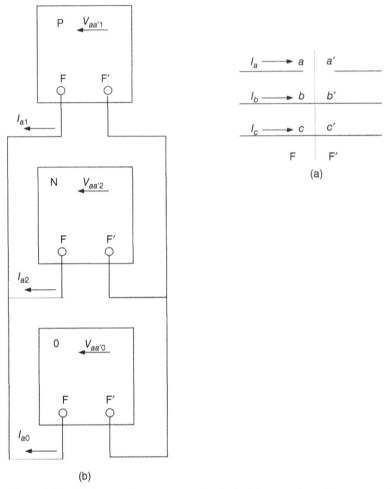

Figure 6.14 (a) One conductor open series fault; (b) connection of sequence networks.

Thus,

$$V_{ao1} = V_{ao2} = V_{ao0} = \frac{1}{3}V_{ao}$$

$$I_{a1} + I_{a2} + I_{a0} = 0$$

(6.26)

This suggests that sequence networks are connected in parallel (Figure 6.14b).

Example 6.3 Consider that one conductor is broken on the high-voltage side at the point marked O in Figure 6.10. The equivalent circuit is shown in Figure 6.15.

An induction motor load of 10,000 hp was considered in the calculations for a single line-to-ground fault in Example 2.2. All other *static loads*, that is, lighting and resistance heating loads, were ignored, as these do not contribute to short-circuit currents. Also, all drive system loads, connected through converters are ignored, unless the

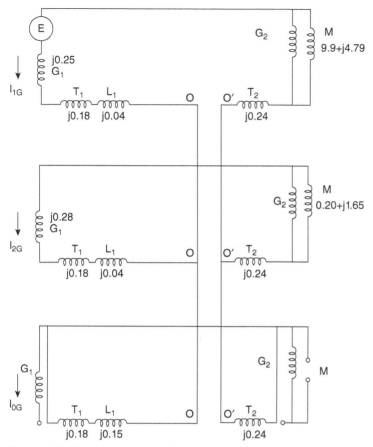

Figure 6.15   Equivalent circuit of an open circuit fault in Example 6.3.

drives are in a regenerative mode. If there are no loads and a broken conductor fault occurs, no load currents flow.

Therefore, for broken conductor faults all loads, irrespective of their types, should be modeled. For simplicity of calculations, again consider that a 10,000-hp induction motor is the only load. Its positive and negative sequence impedances for load modeling will be entirely different from the impedances used in short-circuit calculations.

Chapter 4 discusses induction motor equivalent circuits for positive and negative sequence currents. The range of induction motor impedances per unit (based upon motor kVA base) is

$$X_{1r} = 0.14 - 0.2 < 83° - 75°$$
$$X_{load}^+ = 0.9 - 0.95 < 20° - 26°$$
$$X_{load}^- \approx X_{1r}$$

(6.27)

where X_{1r} = induction motor locked rotor reactance at its rated voltage, X_{load}^{+} = positive sequence load reactance, and X_{load}^{-} = negative sequence load reactance.

The load impedances for the motor are as shown in Figure 6.15. For an open conductor fault as shown in this figure, the load is not interrupted. Under normal operating conditions, the motor load is served by generator G_2, and in the system of Figure 6.15 no current flows in the transmission line L. If an open conductor fault occurs, generator G_2, operating in synchronism, will trip on operation of negative sequence current relays. To proceed with the calculation assume that G_2 is out of service, when the open conductor fault occurs.

The equivalent impedance across an open conductor is

$$[j0.25 + j0.18 + j0.08 + j0.24 + 9.9 + j4.79]_{pos}$$
$$+\{[j0.28 + j0.18 + j0.04 + j0.24 + 0.20 + j1.65]_{neg}$$
$$\text{in parallel with}\quad [j0.18 + j0.15 + j0.24]_{zero}\}$$
$$= 9.918 + j6.771 = 12.0 < 34.32^{0}$$

The motor load current is

$0.089 < -25.84°$ pu (at 0.9 power factor [pf])

The load voltage is assumed as the reference voltage; thus, the generated voltage is

$$V_g = 1 < 0° + (0.089 < -25.84°)(j0.25 + j0.18 + j0.04 + j0.24)$$
$$= 1.0275 + j0.0569 = 1.0291 < 3.17°$$

The positive sequence current is

$$I_{1g} = V_G/Z_t = \left[\frac{1.0291 < 3.17°}{12.00 < 34.32°}\right]$$
$$= 0.0857 < 31.15° = 0.0733 + j0.0443$$

The negative sequence and zero sequence currents are

$$I_{2g} = -I_{1g}\frac{Z_0}{Z_2 + Z_0}$$
$$= (0.0857 < 31.15°)\left[\frac{0.53 < 90°}{2.897 < 86.04°}\right]$$
$$0.0157 < 215.44°$$

$$I_{0g} = -I_{1g}\frac{Z_2}{Z_2 + Z_0}$$
$$= 0.071 < 210.33°$$

Calculate line currents:

$$I_{ag} = I_{1g} + I_{2g} + I_{0g} = 0$$
$$I_{bg} = a^2 I_{1g} + a I_{2g} + I_{0g} = 0.1357 < 250.61°$$
$$I_{cg} = 0.1391 < -8.78°$$

The line currents in the two phases are increased by 52%, indicating serious over-heating. A fully loaded motor will stall. The effect of negative sequence currents in the rotor is simulated by the equation, repeated from Chapter 4:

$$I^2 = I_1^2 + kI_2^2 \tag{6.28}$$

where k can be as high as 6. The motors are disconnected from service by anti-single phasing devices and protective relays. See Chapter 4 for further explanations.

6.8 SHORT-CIRCUIT CALCULATIONS WITH BUS IMPEDANCE MATRIX

Short-circuit calculations using bus impedance matrices apply symmetrical components for the calculations, the logic is identical to that discussed earlier. Consider that the positive, negative, and zero sequence bus impedance matrices Z_{ss}^1, Z_{ss}^2 and Z_{ss}^0 are known and a single line-to-ground fault occurs at the rth bus. The positive sequence current is then injected only at the rth bus and all other currents in the positive sequence current vector are zero. The positive sequence voltage at bus r is given by

$$V_r^1 = -Z_{rr}^1 I_r^1 \tag{6.29}$$

Similarly, the negative and zero sequence voltages are

$$V_r^2 = -Z_{rr}^2 I_r^2$$
$$V_r^0 = -Z_{rr}^0 I_r^0 \tag{6.30}$$

From the sequence network connections for a line-to-ground fault

$$I_r^1 = I_r^2 = I_r^0 = \frac{1.0}{Z_{rr}^1 + Z_{rr}^2 + Z_{rr}^0 + 3Z_f} \tag{6.31}$$

This shows that the following equations can be written for a shorted bus s.

6.8.1 Line-to-Ground Fault

$$I_s^0 = I_s^1 = I_s^2 = \frac{1}{Z_{ss}^1 + Z_{ss}^2 + Z_{ss}^0 + 3Z_f} \tag{6.32}$$

6.8.2 Line-to-Line Fault

$$I_s^1 = -I_s^2 = \frac{1}{Z_{ss}^1 + Z_{ss}^2 + 3Z_f} \tag{6.33}$$

6.8.3 Double Line-to-Ground Fault

$$I_s^1 = \frac{1}{Z_{ss}^1 + \dfrac{Z_{ss}^2 \left(Z_{ss}^0 + 3Z_f\right)}{Z_{ss}^2 + \left(Z_{ss}^0 + 3Z_f\right)}} \tag{6.34}$$

$$I_s^0 = \frac{-Z_{ss}^2}{Z_{ss}^2 + (Z_{ss}^0 + 3Z_f)} I_s^1 \tag{6.35}$$

$$I_s^0 = \frac{-(Z_{ss}^0 + 3Z_f)}{Z_{ss}^2 + (Z_{ss}^0 + 3Z_f)} I_s^1 \tag{6.36}$$

The phase currents are calculated by

$$I_s^{abc} = T_s I_s^{012} \tag{6.37}$$

The voltage at bus j of the system is

$$\begin{vmatrix} V_j^0 \\ V_j^1 \\ V_j^2 \end{vmatrix} = \begin{vmatrix} 0 \\ 1 \\ 0 \end{vmatrix} - \begin{vmatrix} Z_{js}^0 & 0 & 0 \\ 0 & Z_{js}^1 & 0 \\ 0 & 0 & Z_{js}^2 \end{vmatrix} \begin{vmatrix} I_s^0 \\ I_s^1 \\ I_s^2 \end{vmatrix} \tag{6.38}$$

where $j = 1, 2, \ldots, s, \ldots, m$.
   The fault current from bus x to bus y is given by

$$\begin{vmatrix} I_{xy}^0 \\ I_{xy}^1 \\ I_{xy}^2 \end{vmatrix} = \begin{vmatrix} Y_{xy}^0 & 0 & 0 \\ 0 & Y_{xy}^1 & 0 \\ 0 & 0 & Y_{xy}^2 \end{vmatrix} \begin{vmatrix} V_x^0 - V_y^0 \\ V_x^1 - V_y^1 \\ V_x^2 - V_y^2 \end{vmatrix} \tag{6.39}$$

where

$$I_{xy}^0 = \begin{vmatrix} I_{12}^0 \\ I_{13}^0 \\ \cdot \\ I_{mm}^0 \end{vmatrix} \tag{6.40}$$

and

$$\bar{Y}_{xy}^0 = \begin{vmatrix} Y_{12,12}^0 & Y_{12,13}^0 & \cdot & Y_{12,mn}^0 \\ Y_{13,12}^0 & Y_{13,13}^0 & \cdot & Y_{13,mn}^0 \\ \cdot & \cdot & \cdot & \cdot \\ Y_{mn,12}^0 & Y_{mn,13}^0 & \cdot & Y_{mn,mn}^0 \end{vmatrix} \tag{6.41}$$

where Y_{xy}^0 is the inverse of the primitive matrix of the system. Similar expressions apply to positive sequence and negative sequence quantities.
   We will illustrate the procedure with an example.

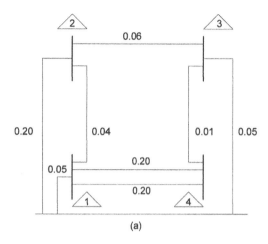

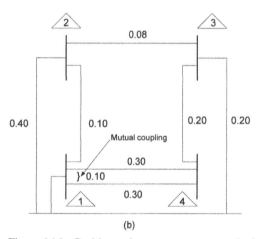

Figure 6.16   Positive and zero sequence networks for Example 6.4.

6.8.4 Calculation Procedure

Example 6.4   Consider a simple system configuration as shown in Figure 6.16a shows the positive and negative sequence network and Figure 6.16b shows the zero sequence network. Both the positive sequence and negative sequence matrices can be reduced to 4×4 matrix, representative of the buses shown in this figure. This calculation of sequence impedance matrices is not documented, see [4–9]:

$$\bar{Z}^{+}, \bar{Z}^{-} = \begin{vmatrix} 0.0321 & 0.0217 & 0.0125 & 0.0223 \\ 0.217 & 0.0397 & 0.0184 & 0.0201 \\ 0.0125 & 0.0184 & 0.0329 & 0.0227 \\ 0.0223 & 0.0201 & 0.0227 & 0.0725 \end{vmatrix}$$

And the zero sequence matrix is

$$
\begin{vmatrix}
0.0184 & 0.0124 & 0.0101 & 0.0142 \\
0.0124 & 0.0665 & 0.0431 & 0.0277 \\
0.0101 & 0.0431 & 0.0782 & 0.0442 \\
0.0142 & 0.0277 & 0.0442 & 0.1292
\end{vmatrix}
$$

A double line-to-ground fault occurs at bus 4 in Figure 6.16. Using the matrices it is required to calculate

- Fault current at bus 4
- Voltage at bus 4
- Voltage at buses 1, 2, and 3
- Fault current flows from buses 3 to 4, 1 to 4, 2 to 3
- Current flow in node 0 to bus 3.

The fault current at bus 4 is first calculated as follows:

$$
\begin{aligned}
I_4^1 &= \frac{1}{Z_{4s}^1 + \dfrac{Z_{4s}^2 \times Z_{4s}^0}{Z_{4s}^2 + Z_{4s}^0}} \\[2mm]
&= \frac{1}{0.0725 + \dfrac{0.0725 \times 0.1292}{0.0725 + 0.1292}} \\[2mm]
&= 8.408 \\[2mm]
I_4^0 &= \frac{-Z_{4s}^2}{Z_{4s}^2 + Z_{4s}^0} I_4^1 \\[2mm]
&= \frac{-0.0725 \times 8.408}{0.0725 + 0.1292} \\[2mm]
&= -3.022 \\[2mm]
I_4^2 &= \frac{-Z_{4s}^0}{Z_{4s}^2 + Z_{4s}^0} I_4^1 \\[2mm]
&= \frac{-0.1292 \times 8.408}{0.0727 + 0.1292} \\[2mm]
&= -5.386
\end{aligned}
$$

The line currents are given by

$$
I_4^{abc} = T_s I_4^{012}
$$

that is,

$$
\begin{vmatrix} I_4^a \\ I_4^b \\ I_4^c \end{vmatrix} = \begin{vmatrix} 1 & 1 & 1 \\ 1 & a^2 & a \\ 1 & a & a^2 \end{vmatrix} \begin{vmatrix} I_4^0 \\ I_4^1 \\ I_4^2 \end{vmatrix}
$$

$$
= \begin{vmatrix} 1 & 1 & 1 \\ 1 & a^2 & a \\ 1 & a & a^2 \end{vmatrix} \begin{vmatrix} -3.022 \\ 8.408 \\ -5.386 \end{vmatrix}
$$

$$
= \begin{vmatrix} 0 \\ -4.533 - j11.946 \\ -4.533 + j11.946 \end{vmatrix} = \begin{vmatrix} 0 \\ 12.777 < 249.2^0 \\ 12.777 < 110.78^0 \end{vmatrix}
$$

Sequence voltages at bus 4 are given by

$$
\begin{vmatrix} V_4^0 \\ V_4^1 \\ V_4^2 \end{vmatrix} = \begin{vmatrix} 0 \\ 1 \\ 0 \end{vmatrix} - \begin{vmatrix} Z_{4s}^0 & & \\ & Z_{4s}^1 & \\ & & Z_{4s}^2 \end{vmatrix} \begin{vmatrix} I_4^0 \\ I_4^1 \\ I_4^2 \end{vmatrix}
$$

$$
= \begin{vmatrix} 0 \\ 1 \\ 0 \end{vmatrix} - \begin{vmatrix} 0.1292 & 0 & 0 \\ 0 & 0.0725 & 0 \\ 0 & 0 & 0.0725 \end{vmatrix} \begin{vmatrix} -3.022 \\ 8.408 \\ -5.386 \end{vmatrix}
$$

$$
= \begin{vmatrix} 0.3904 \\ 0.3904 \\ 0.3904 \end{vmatrix}
$$

Line voltages are, therefore,

$$
\begin{vmatrix} V_4^a \\ V_4^b \\ V_4^c \end{vmatrix} = \begin{vmatrix} 1 & 1 & 1 \\ 1 & a^2 & a \\ 1 & a & a^2 \end{vmatrix} \begin{vmatrix} V_4^0 \\ V_4^1 \\ V_4^2 \end{vmatrix}
$$

$$
= \begin{vmatrix} 1 & 1 & 1 \\ 1 & a^2 & a \\ 1 & a & a^2 \end{vmatrix} \begin{vmatrix} 0.3904 \\ 0.3904 \\ 0.3904 \end{vmatrix} = \begin{vmatrix} 1.182 \\ 0 \\ 0 \end{vmatrix}
$$

Similarly,

$$
\bar{V}_1^{0,1,2} = \begin{vmatrix} 0 \\ 1 \\ 0 \end{vmatrix} - \begin{vmatrix} 0.0142 & 0 & 0 \\ 0 & 0.0223 & 0 \\ 0 & 0 & 0.0223 \end{vmatrix} \begin{vmatrix} -3.022 \\ 8.408 \\ -5.386 \end{vmatrix}
$$

$$
= \begin{vmatrix} 0.0429 \\ 0.8125 \\ 0.1201 \end{vmatrix}
$$

Sequence voltages at buses 2 and 3, similarly calculated, are

$$\bar{V}_2^{0,1,2} = \begin{vmatrix} 0.0837 \\ 0.8310 \\ 0.1083 \end{vmatrix}$$

$$\bar{V}_3^{0,1,2} = \begin{vmatrix} 0.1336 \\ 0.8091 \\ 0.1223 \end{vmatrix}$$

The sequence voltages are converted into line voltages

$$\bar{V}_1^{abc} = \begin{vmatrix} 0.976 < 0^0 \\ 0.734 < 125.2^0 \\ 0.734 < 234.8^0 \end{vmatrix} \quad \bar{V}_2^{abc} = \begin{vmatrix} 1.023 < 0^0 \\ 0.735 < 121.5^0 \\ 0.735 < 238.3^0 \end{vmatrix} \quad \bar{V}_3^{abc} = \begin{vmatrix} 1.065 < 0^0 \\ 0.681 < 119.2^0 \\ 0.681 < 240.8^0 \end{vmatrix}$$

The sequence currents flowing between buses 3 and 4 are given by

$$\begin{vmatrix} I_{34}^0 \\ I_{34}^1 \\ I_{34}^2 \end{vmatrix} = \begin{vmatrix} 1/0.2 & 0 & 0 \\ 0 & 1/0.01 & 0 \\ 0 & 0 & 1/0.01 \end{vmatrix} \begin{vmatrix} 0.1336 - 0.3904 \\ 0.8091 - 0.3904 \\ 0.1223 - 0.3904 \end{vmatrix}$$

$$= \begin{vmatrix} -1.284 \\ 4.1870 \\ -2.681 \end{vmatrix}$$

Similarly, the sequence currents between bus 3 to 2 are

$$\begin{vmatrix} I_{32}^0 \\ I_{32}^1 \\ I_{32}^2 \end{vmatrix} = \begin{vmatrix} 1/0.08 & 0 & 0 \\ 0 & 1/0.06 & 0 \\ 0 & 0 & 1/0.06 \end{vmatrix} \begin{vmatrix} 0.1336 - 0.0837 \\ 0.8091 - 0.8310 \\ 0.1223 - 0.1083 \end{vmatrix} = \begin{vmatrix} 0.624 \\ -0.365 \\ 0.233 \end{vmatrix}$$

This can be transformed into line currents.

$$\bar{I}_{32}^{abc} = \begin{vmatrix} 0.492 \\ 0.69 + j0.518 \\ 0.69 + j0.518 \end{vmatrix} \quad \bar{I}_{34}^{abc} = \begin{vmatrix} 0.222 \\ -2.037 + j5.948 \\ -2.037 + j5.948 \end{vmatrix}$$

The lines between buses 1 and 4 are coupled in the zero sequence network. The $\bar{Y}$ matrix between zero sequence coupled lines is

$$\bar{Y}_{14}^0 = \begin{vmatrix} 3.75 & -1.250 \\ -1.25 & 3.75 \end{vmatrix}$$

Therefore, the sequence currents are given by

$$
\begin{vmatrix} I^0_{14a} \\ I^0_{14b} \\ I^1_{14a} \\ I^1_{14b} \\ I^2_{14a} \\ I^2_{14b} \end{vmatrix} = \begin{vmatrix} 3.75 & -1.25 & 0 & 0 & 0 & 0 \\ -1.25 & 3.75 & 0 & 0 & 0 & 0 \\ 0 & 0 & 5 & 0 & 0 & 0 \\ 0 & 0 & 0 & 5 & 0 & 0 \\ 0 & 0 & 0 & 0 & 5 & 0 \\ 0 & 0 & 0 & 0 & 0 & 5 \end{vmatrix} \begin{vmatrix} 0.0429 - 0.3904 \\ 0.0429 - 0.3904 \\ 0.8125 - 0.3904 \\ 0.8125 - 0.3904 \\ 0.1201 - 0.3904 \\ 0.1201 - 0.3904 \end{vmatrix} = \begin{vmatrix} -0.8688 \\ -0.8688 \\ 2.1105 \\ 2.1105 \\ -1.3515 \\ -1.3515 \end{vmatrix}
$$

Each of the lines carries sequence currents

$$
\bar{I}^{012}_{14a} = \bar{I}^{012}_{14b} = \begin{vmatrix} -0.8688 \\ 2.1105 \\ -1.3515 \end{vmatrix}
$$

Converting into line currents

$$
\bar{I}^{012}_{14a} = \bar{I}^{012}_{14b} = \begin{vmatrix} -0.11 \\ -1.248 - j2.998 \\ -1.248 + j2.998 \end{vmatrix}
$$

Also the line currents between buses 3 and 4 are

$$
\bar{I}^{abc}_{34} = \begin{vmatrix} -0.222 \\ -2.037 - j5.948 \\ -2.037 + j5.948 \end{vmatrix}
$$

Within the accuracy of calculation, the summation of currents (sequence components as well as line currents) at bus 4 is zero. This is a verification of the calculation. Similarly, the vectorial sum of currents at bus 3 should be zero. As the currents between 3 and 4 and 3 and 2 are already known, the currents from node 0 to bus 3 can be calculated.

Concept 6.2   *The bus impedance method as illustrated above may not look simple. The impedance matrix is full matrix and requires storage of each element. The Y-matrix of a large network is very sparse and has a large number of zero elements. In a large system the sparsity may reach 90%, because each bus is connected to only a few other buses. The sparsity techniques are important in matrix manipulation and are not discussed. Some of these matrix techniques are*

- *Triangulation and factorization: Crout's method, bifactorization, and product form.*
- *Solution by forward–backward substitution.*
- *Sparsity and optimal ordering.*

A matrix can be factored into lower, diagonal, and upper form called LDU form. This is of special interest. This formation always requires less computer storage. The sparse techniques exhibit a distinct advantage in computer time required for the solution of a network and can be adapted to system changes, without rebuilding these at every step. All digital computer based programs employ these techniques.

6.9 SYSTEM GROUNDING

The method of symmetrical components is well suited to analysis of system grounding. System grounding refers to the electrical connection between the phase conductors and ground and dictates the manner in which the neutral points of wye connected transformers and generators or artificially derived neutral systems through delta-wye or zig zag transformers are grounded. The equipment grounding refers to the grounding of the exposed metallic parts of the electrical equipment, which can become energized and create a potential to ground, say due to breakdown of insulation or fault, and can be potential safety hazard. The safety of the personnel and human life is of importance. The safety grounding is to establish an equipotential surface in the work area to mitigate shock hazard. The utility systems at high-voltage transmission level, subtransmission level, and distribution level are solidly grounded. The utility generators connected through step-up transformers are invariably high-resistance grounded through a distribution transformer with secondary loading resistor. The industrial systems at medium-voltage level are low-resistance grounded. The implications of system grounding are

- Enough ground fault current should be available to selectively trip the faulty section with minimum disturbance to the system. Ground faults are cleared even faster than the phase faults to limit equipment damage.

- Line-to-ground fault is the most important cause of system temporary overvoltages which dictates the selection of surge arresters.

- Grounding should prevent high overvoltages of ferroresonance, overvoltages of arcing type ground fault and capacitive-inductive resonant couplings. If a generator neutral is left ungrounded, there is possibility of generating high voltages through inductive-capacitive couplings. Ferroresonance can also occur due to presence of generator PTs. In ungrounded systems, a possibility of resonance with high-voltage generation, approaching five times or more of the system voltage exists for values of X_0/X_1 between 0 and -40. For the first phase-to-ground fault, the continuity of operations can be sustained, though unfaulted phases have $\sqrt{3}$ times the normal line-to-ground voltage. All unremoved faults, thus, put greater than normal voltage on system insulation and increased level of conductor and motor insulation may be required. The grounding practices in the industry have withdrawn from this method of grounding.

- The relative magnitude of the fault currents depends upon sequence impedances.

- In industrial systems the continuity of processes is important. The current industry practice is high-resistance grounded systems for low-voltage distributions [9].

Figure 6.17 shows the system grounding methods. A brief discussion follows.

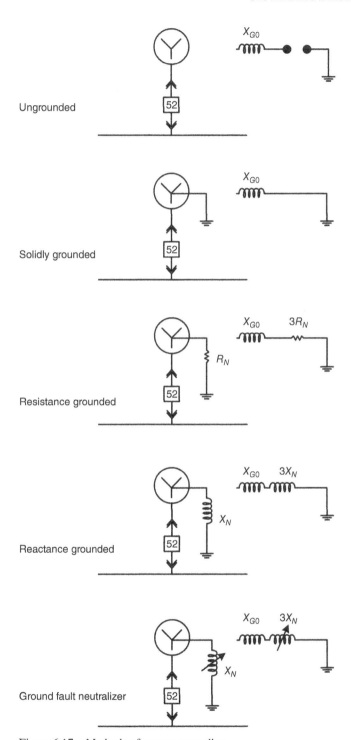

Figure 6.17   Methods of system grounding.

6.9.1 Solidly Grounded Systems

In a solidly grounded system, no intentional impedance is introduced between the system neutral and ground. These systems meet the requirement of "effectively grounded" systems in which the ratio X_0 / X_1 is positive and less than 3.0 and the ratio R_0 / X_0 is positive and less than 1, where X_1, X_0, and R_0 are the positive sequence reactance, zero sequence reactance, and zero sequence resistance, respectively.

The coefficient of grounding (COG) is defined as a ratio of E_{LG} / E_{LL} in percentage, where E_{LG} is the highest rms voltage on a sound phase, at a selected location, during a fault affecting one or more phases to ground, and E_{LL} is the rms phase-to-phase power frequency voltage that is obtained at the same location with the fault removed. Calculations in Example 6.2 show the fault voltage rises on unfaulted phases. Solidly grounded systems are characterized by a COG of 80%.

By contrast, *for ungrounded systems*, definite values cannot be assigned to ratios X_0 / X_1 and R_0/X_0. The ratio X_0 / X_1 is negative and may vary from low to high values. The COG approaches 120%. For values of X_0 / X_1 between 0 and -40, a possibility of resonance with consequent generation of high voltages exists. The overvoltages based on relative values of sequence impedances are plotted in Reference [9]. In an ungrounded system, there is no flow of ground fault current on the first ground fault as we have amply discussed. It was thought that operation can be sustained without immediate fault clearance. However overvoltages of amplitude 4–5 times can occur in these systems due to capacitive inductive couplings. These systems are no longer in use.

The COG affects the selection of rated voltage of the surge arresters and stresses on the insulation systems. Solidly grounded systems are, generally, characterized by COG of 80%. Approximately, a surge arrester with its rated voltage calculated on the basis of the system voltage multiplied by 0.8 can be applied.

The solidly grounded systems have an advantage of providing effective control of overvoltages, which become impressed on or are self-generated in the power system by insulation breakdowns and restriking faults. Yet, these give the highest arc fault current and consequent damage and require immediate isolation of the faulty section. Single line-to-ground fault currents can be higher than the three-phase fault currents [10–12].

Due to high arc fault damage and interruption of processes, the solidly grounded systems are not in much use in the industrial distribution systems. However, ac circuits of less than 50 V and circuits of 50–1000 V for supplying premises wiring systems and single-phase 120/240 V control circuits must be solidly grounded according to NEC [13].

6.9.2 Resistance Grounded Systems

An impedance grounded system has a resistance or reactance connected in the neutral circuit to ground, as shown in Figure 6.17b. In a low-resistance grounded system the resistance in the neutral circuit is so chosen that the ground fault is limited to approximately full load current or even lower, typically 200–400 A. The arc fault damage is reduced, and these systems provide effective control of the overvoltages generated in the system by resonant capacitive-inductive couplings and restriking ground faults. Though the ground fault current is much reduced, it cannot be allowed to be sustained and selective tripping must be provided to isolate the faulty section. For a ground fault current limited to 400 A, the pickup sensitivity of modern ground fault devices can be even lower than 5 A.

The low-resistance grounded systems are adopted at medium voltages, 13.8 kV, 4.16 kV, and 2.4 kV for industrial distribution systems. Also industrial bus connected generators were commonly low-resistance grounded. A recent trend in industrial bus connected medium voltage generator grounding is hybrid grounding systems

6.9.3 High-Resistance Grounded Systems

High-resistance grounded systems limit the ground fault current to a low value, so that an immediate disconnection on occurrence of a ground fault is not required. It is well documented that to control over voltages in the high-resistance grounded systems, the grounding resistor should be so chosen that

$$R_n = \frac{V_{ln}}{3I_c} \qquad (6.42)$$

where V_{ln} is the line to neutral voltage and I_c is the stray capacitance current of each phase conductor. The transients are a minimum when this ratio is unity. This leads to the requirement of accurately calculating the stray capacitance currents in the system[10]. Cables, motors, transformers, surge arresters generators—all contribute to the stray capacitance current. Surge capacitors connected line-to-ground must be considered in the calculations. Once the system stray capacitance is determined, then, the charging current per phase, I_c is given by

$$I_c = \frac{V_{ln}}{X_{c0}} \qquad (6.43)$$

where X_{c0} is the capacitive reactance of each phase, stray capacitance considered lumped together.

This can be illustrated with an example. A high-resistance grounding system for a wye-connected neutral of a 13.8 kV–0.48 transformer is shown in Figure 6.18a. This shows that the stray capacitance current per phase of all the distribution system connected to the secondary of the transformer is 0.21A per phase, assumed to be balanced in each phase. Generally for low-voltage distribution systems, a stray capacitance current of 0.1A per MVA of transformer load can be taken, though this rule of thumb is no substitute for accurate calculations of stray capacitance currents. Figure 6.18a shows that under no fault condition, the vector sum of three capacitance currents is zero, as these are 90° displaced with respect to each voltage vector and therefore 120° displaced with respect to each other Thus, the grounded neutral does not carry any current and the neutral of the system is held at the ground potential, Figure 6.18b. As,

$$I_{c1} + I_{c2} + I_{c3} = 0 \qquad (6.44)$$

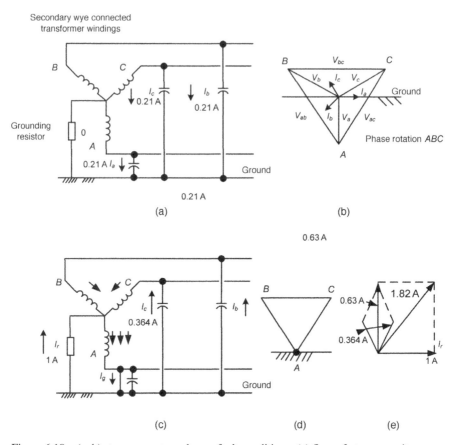

Figure 6.18   (a, b) stray currents under no-fault conditions; (c) flow of stray capacitance current and resistor current for a single line-to-ground fault in phase a; (d) voltage to ground; (e) summation of resistive and capacitive currents.

no capacitance current flows into the ground. On occurrence of a ground fault, say in phase a, the situation is depicted in Figure 6.18c and 6.18d. The capacitance current of faulted a phase is short-circuited to ground. The faulted phase, assuming zero fault resistance is at the ground potential, Figure 6.18d and the other two phases have line-to-line voltages with respect to ground. Therefore, the capacitance current of the unfaulted phases b and c increases proportional to the voltage increase, that is, $\sqrt{3} \times 0.21 = 0.365$ A. Moreover this current in phase b and c reverses, flows through the transformer windings and sums up in the transformer winding of phase a. Figure 6.18e shows that this vector sum $= 0.63$ A.

Now consider that the ground current through the grounding resistor is limited to 1 A only. This is acceptable according to Equation (6.42) as the stray capacitance current is 0.63 A. This resistor ground current also flows through transformer phase winding a to the fault and the total ground fault current is $I_g = \sqrt{1^2 + 0.63^2} = 1.182$ A, Figure 6.18e.

The above analysis assumes a full neutral shift, ignores the fault impedance itself and assumes that the ground grid resistance and the system zero sequence impedances are zero. Practically, the neutral shift will vary.

Though immediate shutdown is prevented, the fault situation should not be prolonged; the fault should be localized and removed. There are three reasons for this.

1. Figure 6.18d shows that the unfaulted phases have voltage rise by a factor of $\sqrt{3}$ to ground. This increases the normal insulation stresses between phase-to-ground. This may be of special concern for low-voltage cables. If the time required to de-energize the system is indefinite, 173% insulation level for the cables must be selected. However, NEC does not specify 173% insulation level and for 600 V cables insulation levels correspond to 100% and 133%. Also Reference [6] specifies that the actual operating voltage on cables should not exceed 5% during continuous operation and 10% during emergencies. This is of importance when 600-V nominal three-phase systems are used for power distributions. The dc loads served through 6-pulse converter systems will have a dc voltage of 648 V and 810 V, respectively, for 480-V and 600-V rms ac systems.

2. Low levels of fault currents if sustained for long time may cause irreparable damage. Though the burning rate is slow, but the heat energy released over the course of time can damage cores and windings of rotating machines even for ground currents as low as 3–4 A. This has been demonstrated in test conditions.

3. A first ground fault left in the system increases the probability of a second ground fault on other phase. If this happens, then it amounts to a 2-phase to ground fault with some interconnecting impedance depending upon the fault location. The potentiality of equipment damage and burnout increases, References [14–17].

6.9.4 Coefficient of Grounding

Symmetrical components can be applied for rigorous calculations of COG. Also simplified equations can be applied for calculation of COG. Sometimes we define EFF (IEC standards, earth fault factor). It is simply

$$EFF = \sqrt{3}COG \tag{6.45}$$

COG can be calculated by the equations described below and more rigorously by the sequence component matrix methods as illustrated above.

Single line-to-ground fault:

$$COG(\text{phase } b) = -\frac{1}{2}\left(\frac{\sqrt{3}k}{2+k}+j1\right)$$

$$COG(\text{phase } c) = -\frac{1}{2}\left(\frac{\sqrt{3}k}{2+k}-j1\right) \tag{6.46}$$

Double line-to-ground fault:

$$\text{COG(phase } a) = \frac{\sqrt{3}k}{1 + 2k} \tag{6.47}$$

where k is given by

$$k = \frac{Z_0}{Z_1} \tag{6.48}$$

To take into account of fault resistance, k is modified as follows:

Single line-to-ground fault:

$$k = \left(R_0 + R_f + jX_0\right) / (R_1 + R_f + jX_1) \tag{6.49}$$

For double line-to-ground fault:

$$k = \left(R_0 + 2R_f + jX_0\right) / (R_1 + 2R_f + jX_1) \tag{6.50}$$

If R_0 and R_1 are zero, then the above equations reduce to

For single line-to-ground fault

$$\text{COG} = \frac{\sqrt{k^2 + k + 1}}{k + 2} \tag{6.51}$$

For double line-to-ground fault

$$\text{COG} = \frac{\sqrt{3}k}{2k + 1} \tag{6.52}$$
$$\text{Where } k \text{ is now} = X_0/X_1 \tag{6.53}$$

In general, fault resistance will reduce COG, except in low-resistance systems.

Example 6.5   Calculate the COG at the faulted bus B in Example 6.2. Then calculate COG if generator G_2 is grounded through a 400 A resistor.

In Example 6.2, all resistances are ignored. A voltage of 1.034 pu was calculated on the unfaulted phases, which gives a COG of 0.597. This is low, because all resistances have been neglected.

If the generator is grounded through 400 A resistor, then $R_0 = 19.19$ ohms, the positive sequence reactance is 0.4 ohms, and the zero sequence reactance is 0.38 ohms, which is much smaller than R_0. *In fact, in a resistance grounded or high-resistance grounded system, the sequence components are relatively small and the ground fault current can be calculated based upon the grounding resistor alone.* The total ground fault current will reduce to approximately 400 A. This gives a COG of approximately 100%. This means that phase-to-ground voltage on unfaulted phases will be equal to line-to-line voltage.

Concept 6.3 *The calculations of short-circuit currents and overvoltages due to faults go hand-in-hand. The application of symmetrical components for these calculations is demonstrated in this chapter. Reference [9] provides quick estimation curves of overvoltages based on R_0/X_1 and X_2/X_1.*

REFERENCES

[1] Transformer Connections (Including Auto-transformer Connections). Publication no. GET-2H. Pittsfield, MA: General Electric, 1967.

[2] ANSI/IEEE Std. C57.12.00-2006, General Requirements of Liquid-Immersed Distribution, Power, and Regulating Transformers.

[3] ANSI Std. C57.12.70-1978, Terminal Markings and Connections for Distribution and Power Transformers.

[4] PL Corbeiller. *Matrix Analysis of Electrical Networks*. Cambridge, MA: Harvard University Press, 1950.

[5] WE Lewis and DG Pryce. *The Application of Matrix Theory to Electrical Engineering*. London: E&FN. Spon, 1965.

[6] HE Brown. *Solution of Large Networks by Matrix Methods*. New York: Wiley Interscience, 1975.

[7] SA Stignant. *Matrix and Tensor Analysis in Electrical Network Theory*. London: Macdonald, 1964.

[8] RB Shipley. *Introduction to Matrices and Power Systems*. New York: Wiley, 1976.

[9] *Electrical Transmission and Distribution Reference Book*, 4th edition. East Pittsburgh, PA: Westinghouse Electric Corp., 1964.

[10] HL Stanback. Predicting damage from 277 volt single-phase-to-ground arcing faults. *IEEE Transactions on Industry Applications*, vol. 13, no. 4, pp. 307–314, July/August 1977.

[11] RH Kaufman and JC Page. Arcing fault protection for low voltage power distribution systems—nature of the problem. *IEEE Transactions on Industry Applications*, vol. 79, pp. 160–167, June 1960.

[12] NEMA PB1-2. Application Guide for Ground Fault Protective Devices for Equipment, 1977.

[13] ANSI/NFPA 70. National Electric Code, 2008.

[14] JC Das. Ground fault protection of bus-connected generators in an interconnected 13.8 kV system. *IEEE Transactions on Industry Applications*, vol. 43, pp. 453–461, no. 2, March/April 2007.

[15] DS Baker. Charging current data for guess work-free design of high resistance grounded systems. *IEEE Transactions on Industry Applications*, vol. IA-15, no. 2, pp. 136–140, March/April 1979.

[16] ICEA Pub. S-61-40, NEMA WCS, Thermoplastic Insulated Wire for Transmission and Distribution of Electrical Energy, 1979.

[17] JR Dunki-Jacobs. The reality of high resistance grounding. *IEEE Transactions on Industry Applications*, vol. IA-13, pp. 469–475, September/October 1977.

FURTHER READING

Blackburn, JL. *Symmetrical Components for Power Systems Engineering*. New York: Marcel Dekker, 1993.

Calabrase, GO. *Symmetrical Components Applied to Electric Power Networks*. New York: Ronald Press Group, 1959.

Das, JC. Grounding of AC and DC low-voltage and medium-voltage drive systems. *IEEE Transactions on Industry Applications*, vol. 34, no. 1. pp. 295–216, January/February 1998.

Gross, CA. *Power System Analysis*. New York: John Wiley & Sons, 1979.

IEEE Std. 142. IEEE Recommended Practice for Grounding of Industrial and Commercial Power Systems, 1991.

Myatt, LJ. *Symmetrical Components*. Oxford/London: Paragon Press, 1968.

Robertson, WF, and Das, JC. Grounding medium voltage mobile or portable equipment. *Industry Applications Magazine*, vol. 6, no. 3, pp. 33–42, May/June 2000.

Smith, DR. Digital simulation of simultaneous unbalances involving open and faulted conductors. *IEEE Transactions PAS*, vol. 89, no. 8, pp. 1826–1835, 1970.

Stevenson, WD. *Elements of Power System Analysis*, 4th edition. New York: McGraw-Hill, 1982.

CHAPTER 7

SOME LIMITATIONS OF SYMMETRICAL COMPONENTS

IN **CHAPTER 1**, we discussed that the method of symmetrical components is applicable to balanced systems. *The symmetrical portion of the network is considered to be isolated, to which an unbalanced condition is applied at the unbalance point. Practically, the power systems are not perfectly balanced and some asymmetry always exists.*

Normally, three-phase systems can be considered balanced. Though some unbalance may exist due to asymmetry in transmission lines, machine impedances, and system voltages, yet these are often small and may be neglected. A single-phase positive sequence network of a three-phase system is adequate for balanced systems, unbalance voltages, and unbalance in three-phase networks can also occur simultaneously. The unbalances cannot be ignored in every case, that is, a distribution system may serve considerable single-phase loads. In such cases, three-phase models are required. A three-phase network can be represented both in impedance and admittance form. The matrix methods for network solution, the primitive network, formation of loop and bus impedance and admittance matrices, and transformations are discussed in References [1–5].

For a three-phase network, the power flow equations can be written as

$$\left| \begin{matrix} V_{pq}^a \\ V_{pq}^b \\ V_{pq}^c \end{matrix} \right| + \left| \begin{matrix} e_{pq}^a \\ e_{pq}^b \\ e_{pq}^c \end{matrix} \right| = \left| \begin{matrix} Z_{pq}^{aa} & Z_{pq}^{ab} & Z_{pq}^{ac} \\ Z_{pq}^{ba} & Z_{pq}^{bb} & Z_{pq}^{bc} \\ Z_{pq}^{ca} & Z_{pq}^{cb} & Z_{pq}^{cc} \end{matrix} \right| \left| \begin{matrix} i_{pq}^a \\ i_{pq}^b \\ i_{pq}^c \end{matrix} \right| \tag{7.1}$$

The equivalent three-phase circuit is shown in Figure 7.1a, and its single line representation in Figure 7.1b. In the condensed form, Equation (7.1) is

$$V_{pq}^{abc} + e_{pq}^{abc} = Z_{pq}^{abc} i_{pq}^{abc} \tag{7.2}$$

Similarly, for a three-phase system, in the admittance form:

$$\left| \begin{matrix} i_{pq}^a \\ i_{pq}^b \\ i_{pq}^c \end{matrix} \right| + \left| \begin{matrix} j_{pq}^a \\ j_{pq}^b \\ j_{pq}^c \end{matrix} \right| = \left| \begin{matrix} y_{pq}^{aa} & y_{pq}^{ab} & y_{pq}^{ac} \\ y_{pq}^{ba} & y_{pq}^{bb} & y_{pq}^{bc} \\ y_{pq}^{ca} & y_{pq}^{cb} & y_{pq}^{cc} \end{matrix} \right| \left| \begin{matrix} V_{pq}^a \\ V_{pq}^b \\ V_{pq}^c \end{matrix} \right| \tag{7.3}$$

Understanding Symmetrical Components for Power System Modeling, First Edition. J.C. Das.
© 2017 by The Institute of Electrical and Electronics Engineers, Inc. Published 2017 by John Wiley & Sons, Inc.

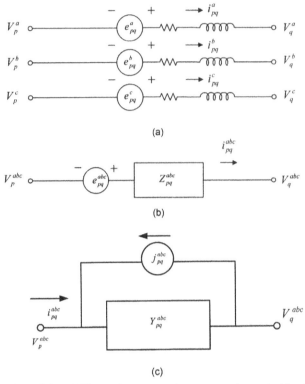

Figure 7.1 (a) Three-phase network representation, primitive impedance matrix; (b, c) single-line representation of three-phase network.

In the condensed form, we can write

$$i_{pq}^{abc} + j_{pq}^{abc} = y_{pq}^{abc} V_{pq}^{abc} \qquad (7.4)$$

This is shown in Fig 7.1 (c) A three-phase load flow study is handled much like a single-phase load flow. Each voltage, current, and power becomes a three-element vector and each single-phase admittance element is replaced by a 3 × 3 admittance matrix.

7.1 PHASE COORDINATE METHOD

As stated before, the assumptions of a symmetrical system are not valid when the system is unbalanced. Untransposed transmission lines, large single-phase traction loads, and bundled conductors are some examples. Unbalanced currents and voltages can give rise to serious problems in the power system, that is, negative sequence currents have a derating effect on generators and motors (Chapter 4) and ground currents can increase the coupling between transmission line conductors. Where the systems are initially coupled, then, even after symmetrical component transformation the equations remain coupled. By representing the system *in phase co-ordinates, that is, phase voltages, currents, impedances, or admittances, the initial physical identity of the system is maintained.* Using the system in the *phase frame of reference*, a generalized analysis of the power system network can be developed for unbalance, that is, short-circuit or load flow conditions [6–8]. The method uses a nodal *Y* admittance matrix and, due to its sparsity, optimal ordering techniques are possible. Series and shunt faults and multiple unbalanced faults can be analyzed. The disadvantage is that it takes more iterations to arrive at a solution.

Transmission lines, synchronous machines, induction motors, and transformers are represented in greater detail. The solution technique can be described in the following steps:

- The system is represented in phase-frame of reference.
- The nodal admittance matrix is assembled and modified for any changes in the system.
- The nodal equations are formed for solution.

The nodal admittance equation is the same as for a single-phase system:

$$\bar{Y}\bar{V} = \bar{I} \qquad (7.5)$$

Each node is replaced by three equivalent separate nodes. Each voltage and current is replaced by phase-to-ground voltages and three-phase currents; I and V are column vectors of nodal phase currents and voltages. Each element of $\bar{Y}$ is replaced with a 3×3 nodal admittance sub-matrix. Active sources such as synchronous machines can be modeled with a voltage source in series with passive elements. Similarly, transformers, transmission lines, and loads are represented on a three-phase basis. The system Y matrix is modified for the conditions under study, for example, a series fault on opening a conductor can be simulated by Y-matrix modification. The shunt faults, that is, single phase-to-ground, three phase-to-ground, two phase-to-ground, and their combinations can be analyzed by the principal of super-imposition.

Consider a phase-to-ground fault at node k in a power system. It is equivalent to setting up a voltage V_f at k equal in magnitude but opposite in sign to the prefault voltage of the node k. The only change in the power system that occurs due to fault may be visualized as the application of a fault voltage V_f at k and the point of zero potential. If the effect of V_f is superimposed upon the prefault state, the fault state can be analyzed. To account for effect of V_f all emf sources are replaced by their internal admittances and converted into equivalent admittance based on the prefault nodal voltage. Then, from Equation (7.5),

$$I_i = 0, \quad \text{and} \quad V_k = \text{prefault voltage}$$
$$i = 1, 2, \ldots, N, \qquad i \neq k \qquad (7.6)$$

The fault current is

$$i_k = \sum_{i=1}^{i=N} Y_{ki} E_i \qquad (7.7)$$

where E_i is the net *postfault* voltage.

For two single line-to-ground faults occurring at two different nodes p and q:

$$I_i = 0, \quad i = 1, 2, \dots, N$$
$$i \neq p, q \tag{7.8}$$

where V_p and V_q are equal to prefault values. Nodes p and q may represent any phase at any busbar. The currents I_p and I_q are calculated from

$$I_k = \sum_{i=1}^{N} Y_{ki} E_i \quad k = p, q \tag{7.9}$$

Thus, calculation of multiple unbalanced faults is as easy as a single line-to-ground fault, which is not the case with the symmetrical component method.

7.2 THREE-PHASE MODELS

Three-phase models of transformers and conductors are described in Chapter 5.

7.2.1 Generators

The generators can be modeled by an internal voltage behind the generator transient reactance. This model is different from the power flow model of a generator, which is specified with a power output and bus voltage magnitude.

The positive, negative, and zero sequence admittances of a generator are well identified. The zero sequence admittance is

$$Y_0 = \frac{1}{R_0 + jX_0 + 3(R_g + jX_g)} \tag{7.10}$$

where R_0 and X_0 are the generator zero sequence resistance and reactance and R_g and X_g are the resistance and reactance added in the neutral grounding circuit; R_g and X_g are zero for a solidly grounded generator. Similarly,

$$Y_1 = \frac{1}{X'_d} \tag{7.11}$$

$$Y_2 = \frac{1}{X_2} \tag{7.12}$$

where X'_d is the generator direct axis transient reactance and X_2 is the generator negative sequence reactance (resistances ignored). These sequence quantities can be related to the phase quantities as follows:

$$
\bar{Y}^{abc} = \begin{vmatrix} Y_{11} & Y_{12} & Y_{13} \\ Y_{21} & Y_{22} & Y_{23} \\ Y_{31} & Y_{32} & Y_{33} \end{vmatrix}
$$

$$
= \frac{1}{3} \bar{T}_s \bar{Y}^{012} \bar{T}_s^t = \begin{vmatrix} Y_0 + Y_1 + Y_2 & Y_0 + aY_1 + a^2Y_2 & Y_0 + a^2Y_1 + aY_2 \\ Y_0 + a^2Y_1 + aY_2 & Y_0 + Y_1 + Y_2 & Y_0 + aY_1 + a^2Y_2 \\ Y_0 + aY_1 + a^2Y_2 & Y_0 + a^2Y_1 + aY_2 & Y_0 + Y_1 + Y_2 \end{vmatrix}
$$

$$\tag{7.13}$$

where a is vector operator $1 < 120°$.

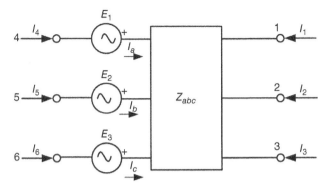

Figure 7.2   Circuit of a three-phase element which can represent a generator.

The following treatment is based on Reference [8]. Figure 7.2 shows a general three-phase circuit element containing sources of EMF in each phase and self- and mutual impedances between phases, which can be summarized by the matrix elements of series impedance matrix Z_{abc}. For the current and voltage reference directions shown, the nodal equations of the current injected into nodes are

$$
\begin{vmatrix} I_1 \\ I_2 \\ I_3 \\ I_4 \\ I_5 \\ I_6 \end{vmatrix} = \begin{vmatrix} Y_{abc} + Y_{123,shunt} & \vdots & -Y_{abc} \\ \cdots & \vdots & \cdots \\ -Y_{abc} & \vdots & Y_{abc} + Y_{456,shunt} \end{vmatrix} \begin{vmatrix} V_1 \\ V_2 \\ V_3 \\ V_4 \\ V_5 \\ V_6 \end{vmatrix} + \begin{vmatrix} -Y_{abc} & \vdots & \\ \cdots & \vdots & \cdots \\ & \vdots & Y_{abc} \end{vmatrix} \begin{vmatrix} E_1 \\ E_2 \\ E_3 \\ E_4 \\ E_5 \\ E_6 \end{vmatrix}
$$
(7.14)

If any nodes are short-circuited the equation can be modified. If nodes 4, 5, 6 are joined together to form a neutral point, then the 6 equations in Equation 7.14 will be reduced to 4 by summation of rows 4, 5, and 6; $I_N = I_4 + I_5 + I_6$ and summation of columns $V_N = V_4 + V_5 + V_6$.

Then from Equations (7.13) and (7.14), the equations for a wye-connected generator with neutral N are

$$
\begin{vmatrix} I_1 \\ I_2 \\ I_3 \\ I_N \end{vmatrix} = \begin{vmatrix} Y_{11} & Y_{12} & Y_{13} & -Y_0 \\ Y_{21} & Y_{22} & Y_{23} & -Y_0 \\ Y_{31} & Y_{32} & Y_{33} & -Y_0 \\ -Y_0 & -Y_0 & -Y_0 & -3Y_0 \end{vmatrix} \begin{vmatrix} V_1 \\ V_2 \\ V_3 \\ V_N \end{vmatrix} + \begin{vmatrix} -Y_{11} & -Y_{12} & -Y_{13} & 0 \\ -Y_{21} & -Y_{22} & -Y_{23} & 0 \\ -Y_{31} & -Y_{32} & -Y_{33} & 0 \\ 0 & 0 & 0 & Y_0 \end{vmatrix} \begin{vmatrix} E_1 \\ E_2 \\ E_3 \\ E_N \end{vmatrix}
$$
(7.15)

where, for balanced EMFs per phase,

$$
\begin{aligned}
E_1 &= E_a, E_2 = E_b, E_3 = E_c \\
E_N &= E_1 + E_2 + E_3 = 0
\end{aligned}
$$
(7.16)

If the neutral is grounded through an admittance Y_N, Y_N appears as a normal shunt impedance term in element Y_{44} or Y_{NN} which becomes $(3Y_0 + Y_N)$.

Referring to Figure 7.2, we can write

$$I_a = S_1^* / E_1^* \quad I_b = S_2^* / E_2^* \quad I_c = S_c^* / E_3^* \tag{7.17}$$

$$I_a = S_1^* / E_1^* \quad I_b = S_2^* / aE_1^* \quad I_c = S_c^* / a^2 E_1^* \tag{7.18}$$

$$I_a + I_b + I_c = \frac{S^*}{E_1^*} = \frac{S_1^* + S_2^* + S_3^*}{E_1^*} \tag{7.19}$$

where S_1, S_2 and S_3 are the individual phase powers, S is the total power, and E_1 is the positive sequence voltage behind the transient reactance. For a solidly grounded system the neutral voltage is zero. The internal machine voltages E_1, E_2, and E_3 are balanced; however, the terminal voltages V_1, V_2, and V_3 depend on internal machine impedances and unbalance in machine currents, I_a, I_b, and I_c. Because of unbalance, each phase power is not equal to one-third of the total power. I_1, I_2, and I_3 are injected currents and I_n is the neutral current. Equation (7.15) can model unbalances in the machine inductances and external circuit [2].

7.2.2 Generator Model for Cogeneration

The co-generators in distribution system load flow are not modeled as PV type machines, that is, to control the bus voltage. They are controlled to maintain a constant power and power factor, and power factor controllers may be required. Thus, for load flow, the synchronous generators can be modeled as constant complex power devices. The induction generators require reactive power which will vary with the terminal voltage. Assuming a voltage close to rated voltage these can also be modeled as P-Q devices. Figure 7.3a shows the Norton equivalent of the generator and Figure 7.3b shows the load flow calculation procedure [4]. The generator is represented by three injected currents. For the short-circuit calculations, the generator model is the same, except that I_1 is kept constant, as the internal voltage of the generator can be assumed not to change immediately after the fault.

The energy system research center (ESRC) at the University of Texas at Arlington has developed a program "GDAS" for unbalance load flow in the distribution systems [9].

7.2.3 Load Models

Based on test data, a detailed load model can be derived, and the voltage/current characteristics of the models are considered. The models are not entirely three-phase balanced types and single-phase loads give rise to unbalances. A load window can be first constructed and percent of each load type allocated (Figure 7.4). By testing, the power/voltage characteristics of most load types is known. As an example, in fluorescent lighting the power requirement reduces when the voltage dips and increases as the voltage is restored to the operating voltage. Conversely for air-conditioning loads, the power requirement will increase as the voltage rises and also as the voltage dips below rated, giving U-shaped curves. A typical three-phase load is shown in Figure 7.5. The unbalance is allowed by load current injections.

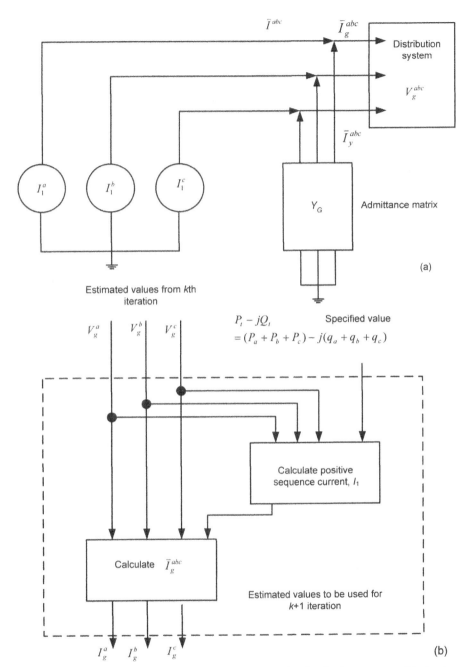

Figure 7.3   (a) Norton equivalent circuit of a generator for distribution systems; (b) circuit for load flow calculations.

Incandescent lighting	Fluorescent lighting	Space heating	Dryer	Refrig. freezer	Elect. range	TV	Others	Total =100%

The width of the window shown is
not representative of the % load

Figure 7.4   Representation of a load window.

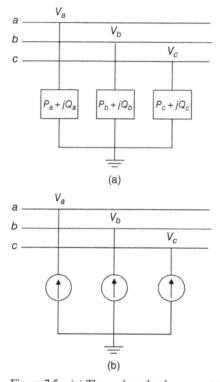

(a)

(b)

Figure 7.5   (a) Three-phase load representation; (b) equivalent circuit injection.

7.3 MULTIPLE GROUNDED SYSTEMS

In Chapter 6, we calculated the overvoltages for single-line-to-ground fault and COG. However for multiple grounded systems, problems occur with symmetrical component method.

Figure 7.6 shows typical grounding practice for wye service entrance served by a wye multiple-grounded medium voltage system in the North America. Note the multiple grounds of the neutral conductor (PEN—protected neutral). The practice of grounding of commercial and residential facilities in the United States requires that

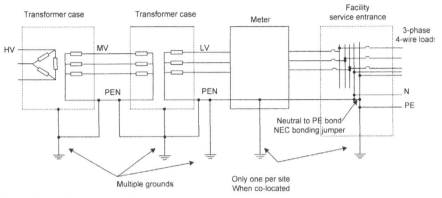

Figure 7.6   Typical grounding practice for wye-service entrance served by multiple grounded, medium voltage system in North American systems.

the neutral conductor is bonded to the ground conductor at the service entrance, and both are bonded to the building ground. There cannot be N-G surge at the service entrance. However, L-N surges within the building can produce N-G surges at the end of a branch circuit.

Further implications of multiple grounded distribution systems are shown in Figure 7.6. NESC (National Electric Safety Code) requires that the neutral on multiple grounded wye distribution systems have a minimum of four earth connections per mile. This also applies to direct buried underground cables. The voltage between neutral and earth can originate from variety of sources. A 60-Hz voltage can exist between objects connected to neutral and earth. A short-duration transient can exist, when the lightning current is dissipated into the earth. A differential voltage between the neutral and ground is more likely to occur when the same service transformer feeds two or more consumers.

Figure 7.7 shows the grounding practice for industrial establishments. Here the neutral is grounded only at one point, at the source. This figure shows that the neutral

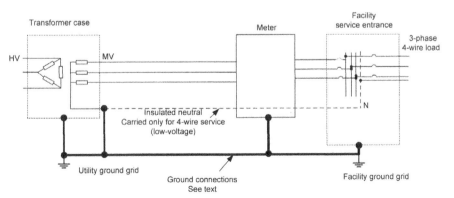

Figure 7.7   Typical grounding practice for industrial systems, the transformer neutral is only grounded at the source in North American Systems.

from the utility transformer is not required to be run for industrial plant medium voltage three-phase loads. In case the industrial plant needs some loads like lighting and controls to be served from low-voltage grounded systems, these lower voltages are served from a separate transformer with artificially derived neutral. In case the service is at low voltage, a neutral may be run to supply phase-to-neutral loads, *but it is* not grounded anywhere in the plant except at the service transformer. There is no bonding of neutral conductor with the ground at the service entrance, a practice which is invariably followed for industrial medium or high-voltage grounded systems or separately derived industrial systems.

7.3.1 Equivalent Circuit of Multiple Grounded Systems

Figure 7.6 of a multiple grounded system shows that the grounds at various points cannot be at the same potential. This figure shows the current flow in the multiple grounded neutral under normal operation. The load current flows through line to neutral, but as the neutral is grounded at the consumer premises (ground CG) and also at multiple points the neutral current returns to the utility transformer through multiple paths, the sharing of current depends upon the relative impedances on the grounding circuit. An equivalent impedance diagram is shown in Figure 7.8

The symmetrical component method of analysis can be based on the assumption that the grounding conditions are uniform along the entire length, but it cannot analyze the effect of individual grounding electrodes. A serious error occurs when the line runs through areas of high resistivity.

7.3.2 Equivalent Circuit Approach

The system may be analyzed using equivalent circuit concepts. The variations of electrode resistances and uneven spans can be easily accommodated. The current and voltage equations in the network form a set of simultaneous operations which can be

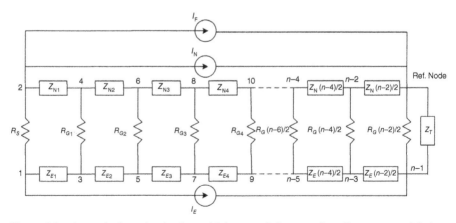

Figure 7.8   An equivalent circuit of a multiple grounded system for a line-to-ground fault at the remoter node.

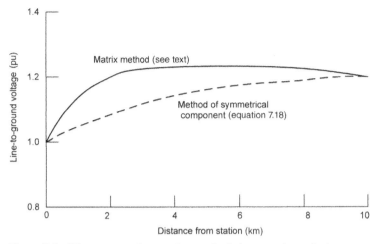

Figure 7.9   Line-to-ground overvoltage calculation, matrix method verses symmetrical component method; see text.

solved by matrix manipulations and a solution of voltages and currents in the entire system can be obtained. A detailed computer program known as "VOLTPROF" has been developed by Ontario Hydro for this purpose. Figure 7.9 shows the comparative results of simulation for a particular location. For the symmetrical component calculations it is assumed that the earth and neutral are conductively coupled through an impedance of negligible resistance. Then the zero sequence impedance can be written as

$$Z_0 = r_a + r_e = j(x_a + x_e - 2x_d) - \frac{[r_e + j(x_e - 3x_{dn})]^2}{3r_n + r_e + j(3x_n + x_e)} \tag{7.20}$$

where

r_a = Resistance of phase conductor, ohms/mile

x_a = Reactance of phase conductor due to magnetic flux inside one foot radius, ohms /mile

x_d = Reactance of phase conductor due to magnetic flux outside one foot radius, ohms/mile

r_e = earth return path resistance, ohms/mile

x_e = Earth return path reactance, ohms/mile

r_n = resistance of neutral conductor, ohms/mile

x_n = Reactance of neutral conductor due to magnetic flux inside one foot radius, ohms/mile

x_{dn} = Reactance of neutral conductor due to magnetic flux outside one foot radius, ohms/mile

REFERENCES

[1] HE Brown. *Solution of Large Networks by Matrix Methods*. New York: John Wiley & Sons, 1975.

[2] GW Stagg and AH El-Abiad. *Computer Methods in Power Systems Analysis*. New York: McGraw-Hill, 1968.

[3] HE Brown and CE Parson. Short-circuit studies of large systems by the impedance matrix method. *Proc. PICA*. pp. 335–346, 1967.

[4] R Bergen and V Vittal. *Power System Analysis*, 2nd edition. New Jersey: Prentice Hall, 1999.

[5] WJ Maron. *Numerical Analysis*. New York: Macmillan, 1987.

[6] L Roy. Generalized polyphase fault analysis program: calculation of cross country faults. *Proc. IEE (Part B)*, vol. 126, no. 10, pp. 995–1000, 1979.

[7] MA Loughton. The analysis of unbalanced polyphase networks by the method of phase coordinates, Part 1: System representation in phase frame reference. *Proc. IEE*, vol. 115, no. 8, pp. 1163–1172, 1968.

[8] MA Loughton. The analysis of unbalanced polyphase networks by the method of phase coordinates, Part 2: System fault analysis. *Proc. IEE*, vol. 1165, no. 5, pp. 857–865, 1969.

[9] TH Chen. Generlized Distribution System Analysis. PhD Thesis, University of Texas at Arlington, May 1990.

INDEX

ABCD constants (of transmission lines), 40
admittance matrix
    autotransformer-three phase, 29–31
    for generators and unbalance loading, 150
    three-phase conductors, 98–100
    three-phase transformer models, 91–98

balanced system, 1, 147
bus impedance matrix, 131

cable constants, 54
    zero sequence impedance of OH lines and
        cables, 54–56
capacitance of cables, 57
capacitance of lines, 50
    capacitance matrix, 50–53
Carson's Formula, 44–46
    approximations to, 46
characteristic equation, 2
characteristics of sequence networks, 19
characteristics of symmetrical components,
    16–19
Clarke component transformation, 11–12
COG, 140
    calculations of, 143–145
construction of sequence networks, 20–22,
    32–36

decoupling a three-phase symmetrical
        system with symmetrical
        components, 6–8
decoupling a three-phase unbalanced
        system, 10–11
delta-wye transformation of impedances, 36
diagonalization of a matrix, 5

EFF, 143
EMTP, 12

EMTP models of transmission lines,
    58
    frequency dependent model, 60–62
eigenvalues, 2–4
eigenvectors, 2–4

ferroresonance, 138
Fortescue, 1
    originator of sequence component
        theory, 1

generators, *see* synchronous generators
grounding, 138
    COG, 140–141
    EFF, 141
    equipment grounding, 138
    ferroresonance, 138
    high resistance grounding, 141–143
        stray capacitance current, 141
    resistance grounded systems, 140–141
    solidly grounded systems, 140
        NEC requirements, 140
    system grounding, 138–139

harmonics and sequence components, 73,
    87–88
Hermitian matrix,
high resistance grounding, 141–143
    stray capacitance currents, 141

inductance matrix of a generator, 79
induction motors, 81
    equivalent circuit, 81, 83
    harmonic impedances, 84–86
    negative sequence impedance, 83–84
    terminal short-circuit, 86
    zero sequence impedance, 86
instantaneous power theory, 12

Understanding Symmetrical Components for Power System Modeling, First Edition. J.C. Das.
© 2017 by The Institute of Electrical and Electronics Engineers, Inc. Published 2017 by John Wiley & Sons, Inc.

IEEE Press Series
on Power Engineering

Series Editor: M. E. El-Hawary, Dalhousie University, Halifax, Nova Scotia, Canada

The mission of IEEE Press Series on Power Engineering is to publish leading-edge books that cover the broad spectrum of current and forward-looking technologies in this fast-moving area. The series attracts highly acclaimed authors from industry/academia to provide accessible coverage of current and emerging topics in power engineering and allied fields. Our target audience includes the power engineering professional who is interested in enhancing their knowledge and perspective in their areas of interest.